KW-220-792

CAUSEWAY
GCSE
GEOGRAPHY

Frances Cook, Helen Harris, Rachel Lofthouse, Mel Rockett
edited by Mel Rockett

CAUSEWAY PRESS

Contents

1 Population studies

to the student

Many think that population growth is the most important problem facing the world today. Total world population was 5.7 billion in 1995. It is increasing by 90 million people each year - that's three births per second! At these rates, world population will double by 2050.

Most (95 per cent) of this growth is taking place in *less economically developed countries* (LEDCs) where the struggle to overcome poverty is a daily battle. These LEDCs are mainly in Africa, south and east Asia and in South America. Optimists say that humans will find ways of solving this crisis by helping the poor economies of the *South* catch up with the richer *North*. Pessimists believe that, unless this happens soon, problems of famine, disease and war will increase.

In the Enquiry which follows, you will investigate the current rates of population growth in different parts of the world. You will look at the problems and will consider possible solutions.

questions to consider

1 What are the current rates of population growth?
2 Where in the world is population growth the fastest?
3 What are the causes of population growth?
4 What problems does this create?
5 What are the possible solutions?

key ideas

Population growth mainly depends upon the *birth rate* and *death rate* - the number of births or deaths per 1,000 of the population. *Natural increase* is the number of births minus the number of deaths in a population over a given time period.

The *growth rate (or rate of natural increase)* is the rate at which a population is increasing or decreasing in a given year. It is calculated by subtracting the death rate from the birth rate.

Each country has an *optimum population*. This is the level of population, working with available resources, that will bring about the highest standard of living.

In very broad terms, the world can be divided into the *North* and the *South*. The North contains the most economically developed countries (*MEDCs*) and the South contains the least economically developed countries (*LEDCs*).

activities

Using information from this and the facing page:

a) List 2 factors which affect a country's population size.
b) Briefly describe 3 problems that can be caused by rapid population growth.
c) Draw up a table to show the dates (and the number of years) it has taken for the world's population to rise by each billion since 1804.
d) Working with a partner, write down what comes to mind when you think how the growth in world population might affect you.

The Population Clock

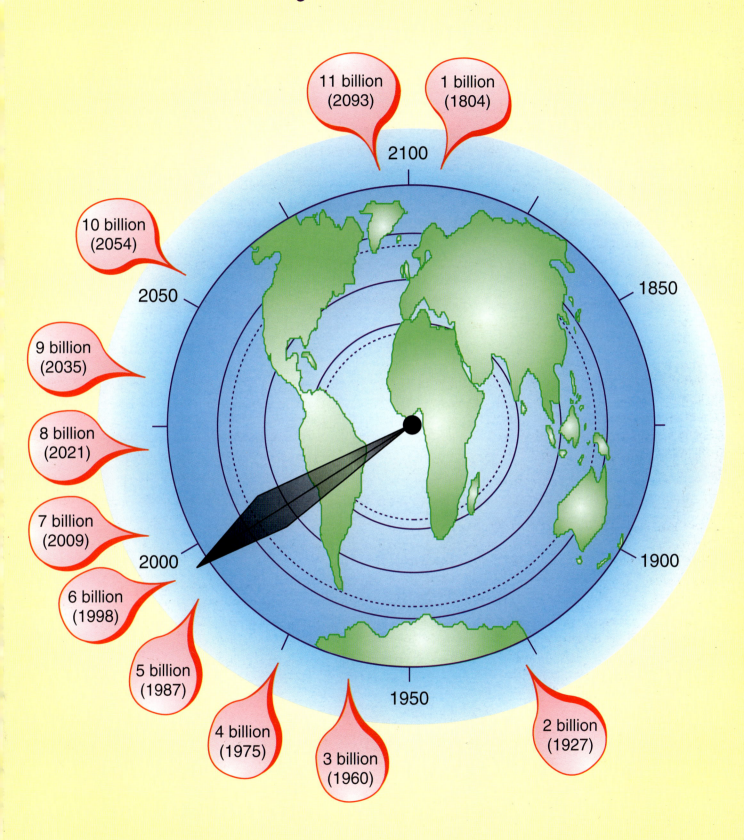

Source: United Nations Population Division

Enquiry

World population growth

In this Enquiry you will study where population is increasing fastest, why this is occurring and what problems are caused by different rates of population growth. Finally, you will look at some possible solutions.

Important global issues such as these are often discussed at world conferences. This is exactly what happened in Cairo, Egypt, at the 1994 International Conference on Population and Development.

The outcome of this Enquiry will be a presentation that you will prepare for such a conference. Your task will be to take the role of a delegate who has been asked to make an introductory presentation that sets out the main world population issues. Before you do this, work through the Stages in the following pages to help you build up your knowledge and understanding.

STAGE 1: What are current population totals and trends?

Over half the world's population lives in Asia. Two Asian countries, China and India, have almost 40 per cent of the world's population. However, at the present time, the population of Africa is rising at a faster rate than that of Asia. Because the largest and fastest growing populations are in countries with low average incomes, there is an ever present threat of hunger and famine.

Some countries have a higher **population density** than others. Singapore has 2,900 people per square kilometre, Hong Kong has even more - 6,100 people per square kilometre. At the other extreme, Russia has 9 people per square kilometre and Australia has 2 people per square kilometre. If the total world population lived in as crowded conditions as those in Hong Kong, they would fit in an area the size of France and Spain together. So, when considering a country's population and growth rate, it is also useful to consider the population density. Other things being equal, the lower the density, the better able is the country to cope with a rising population.

The resources that follow contain information on world population.

Use the information in the table to draw a graph that shows actual world population growth in the different continents between 1850 and 1990, and projected growth between 1990 and 2025. Write a short description of the information shown on the graph.

Draw up a list of the countries with the ten biggest populations and classify them by continent. Set out your list in the form of a table like that shown in the resources. Then write a summary of the information - where are the countries with the biggest population?

Which of the ten countries have a high (over 200 per sq km), medium (between 20 - 199 per sq km) or low (below 20 per sq km) population density?

Most people in Hong Kong live in high rise apartment blocks. It is one of the most densely populated places in the world.

Continent	Population total (millions)				
	1850	1900	1950	1990	2025
Asia	660	870	1670	3550	4530
Africa	96	120	280	660	1540
South America	30	60	214	560	860
North America	21	66	200	260	340
Europe (incl. former USSR)	270	420	640	820	980
Australasia	0.2	1	15	30	35

Ten countries with the highest populations (1994)

	Population (millions)	Continent	Population density (population per square kilometre)
China	1,190	Asia	124
India	913		278
USA	261		28
Indonesia	190		100
Brazil	159		19
Russia	148		9
Pakistan	126		158
Japan	125		331
Bangladesh	118		819
Nigeria	108		117

Australia has one of the world's lowest population densities.

Population density (1995)

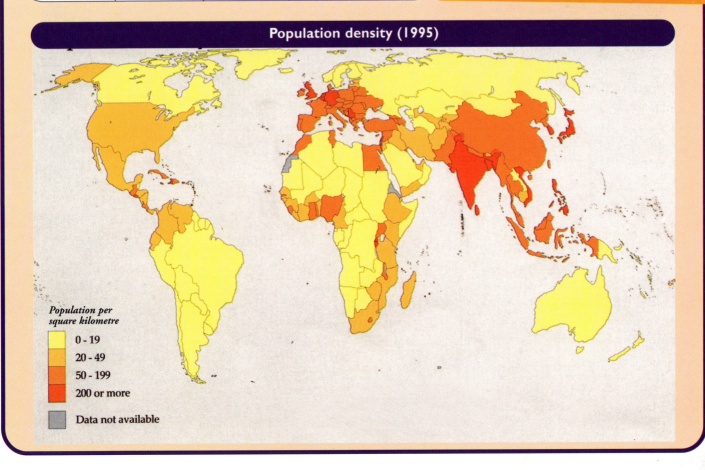

Population per square kilometre

- 0 - 19
- 20 - 49
- 50 - 199
- 200 or more
- Data not available

STAGE 2: Where is population growth the fastest?

The fastest population growth rates are mostly in Africa and in the Middle East. For example, Gambia (in West Africa) had an annual growth rate of 4.1 per cent (this is the same as 41 per 1,000) between 1985 and 1994. Jordan's rate (in the Middle East) was even higher at 5.2 per cent (52 per 1,000). At these rates, Gambia's population would double in 18 years and Jordan's would double in 14 years.

By contrast, the UK's population annual growth rate between 1985 and 1994 was 0.3 per cent (3 per 1,000). At this rate, the population would take 231 years to double. In some parts of Europe, the growth rates are negative. This means that the population is falling. For example, between 1985 and 1994, the annual population change in Hungary was -0.4 per cent and in Portugal was -0.1 per cent.

The growth rate on its own does not tell us how many extra people there will be in the future. For this, we need to know the actual population size as well. A low growth rate for a big population can bring about a much bigger absolute increase than a high growth rate for a small population. For example, Jordan's high annual rate of 5.2 per cent - if it continues - will mean a population rise of 2.8 million in 10 years (Jordan's 1994 population was 4.2 million). On the other hand, China, with a much lower annual growth rate of 1.4 per cent, will have a population increase of over 178 million in the next ten years - if present trends continue. This is because its 1994 population was 1,190 million.

In the resources that follow, you will find information on population growth rates and detailed information on selected countries.

Using the map, briefly describe where the highest and lowest population growth rates are occurring.

Notice that, in the table of statistics on selected countries, the population growth rate is given together with the actual (or absolute) increase experienced in one year.

Which of the countries has the highest growth rate and which has the highest absolute increase? Write a paragraph that explains the difference between the rate of growth and absolute increase.

Use examples to show why the highest population growth rate does not necessarily cause the highest absolute increase.

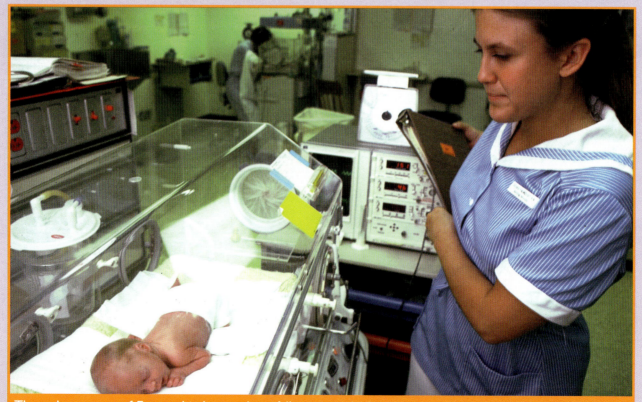

Throughout most of Europe birth rates have fallen so much that they nearly match death rates. This means that the population is barely growing. A negative growth rate means that the birth rate has fallen below the death rate.

Population statistics for selected countries

Country	Total population (millions) 1994	Population growth rate (% per year) 1985 - 94	One year's population growth at 1985 - 94 rate (millions)
Belgium	10.1	0.2	0.02
China	1190.9	1.4	16.67
Ethiopia	53.4	2.9	1.55
Kenya	26.0	2.9	0.75
Japan	124.7	0.4	0.50
Mexico	91.8	2.2	2.02
Thailand	58.7	1.6	0.94
Tanzania	28.8	3.1	0.89
UK	58.0	0.3	0.17
USA	260.5	1.0	2.61

China's population is only growing at half the rate of Kenya's population. But, because the total population is so much bigger, the absolute annual increase in China is over 20 times greater than in Kenya.

Population growth rates (1985-1995)

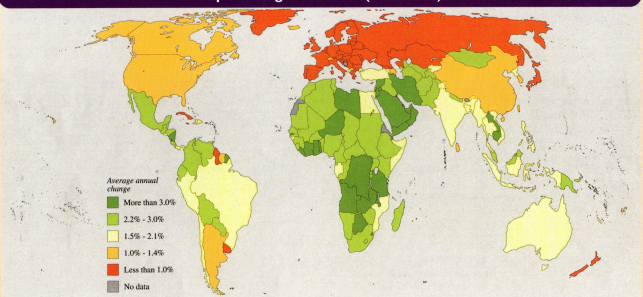

Average annual change

- More than 3.0%
- 2.2% - 3.0%
- 1.5% - 2.1%
- 1.0% - 1.4%
- Less than 1.0%
- No data

The growth rate depends upon the number of births, the number of deaths and migration flows. So, the absolute change in population in one year equals the sum of births and immigrants minus the sum of deaths and emigrants.

The percentage growth rate is calculated by dividing the change in population by the total population, and then multiplying by 100.

STAGE 3: What are the causes of different world trends in population?

In Stage 2, we saw that population is growing fastest in Asia, Africa and South America. These are mainly the poorest parts of the world, with low average incomes and large numbers living at starvation level. The main cause of rapidly rising population in these regions is a falling death rate combined with a high birth rate.

Death rates are falling because of better health care. However, birth rates remain high and this is the big difference between more developed and less developed countries.

The less developed countries are caught in a vicious circle of poverty and rapid population growth. Birth rates are high because people are poor and need children to provide labour and security for old age. If people were better off, they would need fewer children.

In the resources that follow, you are given information on 3 different families. They are from Kenya, Thailand and Belgium. Kenya is a country with a low average income, Thailand is a middle income country and Belgium has a high average income.

Use the information to make notes on why population growth rates differ. In particular:

Why are birth rates different in the three countries? How do women's and other family members' attitudes differ between the countries?

Thailand - a family in Bangkok

Mother: Yuey
Father: Nikorn
Daughter: Keng (11 years)
Son: Pira (5 years)

Yuey

I work for 8 hours a day in a factory that makes radios and cassette players. I need to work to pay for my daughter's education past the compulsory 6th year. Life is not good for those with no education. Today, 85% of children are educated at primary school but I want my daughter to be one of the 28% who go on to secondary school.

Today, life is much better than when I was young. My father was a fisherman on the coast and lost everything in a storm. Our family moved to Bangkok in search of opportunity. Growing up, I worked on a food stall at the docks to pay for my education. My mother, who lives with us, still sells food at a local market.

Women now have more choices and a better education. They are therefore less dependent on having children. More women than ever before are employed - nearly 50% now. Family planning is readily available in both rural and urban areas through district health offices. These are run by the government as part of the 5 Year Development Plan.

I'm glad the future looks better for my children.

Keng

I am glad to be at school. I just hope that I won't be taken out and the money used to continue my brother's education when he is 7. One day I want to become a teacher and help the children in the poorer parts of our city where life is hard. There are many people who are in ill health and not many have an education. One day I would like to get married and have children - but not for a long time yet.

Nikorn

Life has changed a lot in Thailand over the past 20 years. Cities are thriving and growing very rapidly. Our country is popular with tourists and this has helped me get a job as a porter in one of the new skyscraper hotels. I am glad that I can work and help support my family. Yuey and I have decided not to have any more children. Education is so expensive but we think that it is important to give our children a good start in life. We hope that our son and daughter can look after us in our old age.

Yuey's mother is a market trader.

Kenya - a family living in a rural village, 100 km from the nearest large town, Nakuru.

Mother: Maina
Father: Ikaweba
Six children: Atieno, Rose, Muturi, Sarah, Daniel, Isaiah

Maina

I have no formal education. I have lived in this village all my life and my day is filled with farming the land. We grow maize, beans and cassava. I spend my time growing the food to feed my family - weeding, planting, threshing and carrying water. This is typical in our country, nearly all farm work is done by women. I have 6 children, it would have been 7 but my first child died of measles soon after he was born. This would not happen now. Since 1985 all children in the village are vaccinated against this disease and polio. But we still have no safe drinking water in the village and my children are often ill with diarrhoea.

In our society women are regarded as inferior to men, but having children improves our status in the community and means that we are valued by our husbands. If I had an income, it would give me status and power to make my own decisions.

When my husband leaves to find work in Nakuru, I will need my children to help me even more on the farm and in the home. I know about contraception but, you can see, it has been important for me to have children. Anyway, the nearest clinic is in Nakuru which is four hours away on the bus.

Ikaweba

I am head of the household. Men hold the power in this village and it is we who make the decisions regarding our families and the land. Our existence is a difficult one and without new tools and fertiliser the land is becoming less and less productive. Next month I shall travel to Nakuru to try and find work. I do not know when I shall return.
I have no savings, so I need my children to look after me in old age. I hope that at least one of my sons can complete his education and get a good job in the city. Then he can send me money for the rest of the family.

Daniel (11 years)

I work very hard on the farm. I used to help by carrying water and looking after my younger brother but now I am allowed to herd and watch over the cattle. One day I hope to go to school and get a job in the city.

Atieno (19 years)

I am married and expecting my first child in 6 months time. It is common for women to have their first child before 18. I am happy to be married, having a baby outside marriage is unacceptable in our society. There are strict abortion laws which force many women to have illegal, unsafe abortions. I would not want this to happen to me, or my sisters.

With my sisters I still help my mother by carrying water and firewood every day. There is always a lot of work to do on the land, now we are preparing the soil for the next cassava crop. I am worried about how I shall manage to keep working and look after the baby, but I know that the other women in the village will help.

Daniel with two friends.

Belgium - a family living in a Brussels suburb

Monique takes her son to a private nursery.

Monique

I work as a secretary for a Euro MP. In our country everyone has a right to education and, in fact, must attend primary and secondary school. I worked hard to gain the qualifications I needed to become a secretary. I got married when I was 27 and had my son 2 years later. I was very keen to return to work and pursue my career. Nursery provision is excellent in my job and so it made things easier. I don't know if I shall have any more children - it's nice to be able to afford good holidays and all the things we need for our large house.

Although we were both brought up as Catholics, we do not accept the Church's teaching against using birth control. I think that it is important for me to decide if and when I have children.

Mother: Monique
Father: Albert
Son: Pierre (1 year)

Albert

I am a manager for Petrofina, Belgium's largest multinational company. With my work I have to travel a great deal to other parts of Europe and Africa. We enjoy a high standard of living and have pretty well everything we need. Education and health care are excellent in our country. Both Monique's and my parents are still alive, even though mine retired many years ago. They are living in sheltered accommodation quite near to us. We are very lucky when you compare our life with most people in the world.

STAGE 4: How is population growth related to development?

Many people believe that population growth is directly linked to economic development. The best way to tackle rapid population growth and high birth rates is therefore to tackle the problem of poverty. Most developed countries have low birth rates and low death rates. It is clear from the history of these countries that their birth rates fell as their incomes rose.

The way that, for example in the UK, death rates fell and then birth rates fell is described and explained by the **Demographic Transition Model (DTM)**. (A model is a means of explaining and predicting events.) The Demographic Transition Model shows how birth rates, death rates and populations change over time. It is usually shown in graph form and an example is provided in the resources that follow.

Copy the graph of the Demographic Transition Model and the labels that are given underneath it. Mark on the graph where you think Kenya, Thailand and Belgium are at the present time.

Make a written note of which DTM Stage each of the three countries is in and relate this to their income levels. List the difficulties that Kenya faces compared with the other two countries.

Does the information in this Enquiry support the view that 'the best contraception is a high level of income'? Write a short essay on this topic. In your answer make sure that you describe and explain the factors that influence the birth rate. In your opinion, is economic development and a high income level the best means of reducing the birth rate?

Demographic Transition Model

High

Birth and death rates per 1000 people per year

Death rate

Birth rate

Population total

→ Time

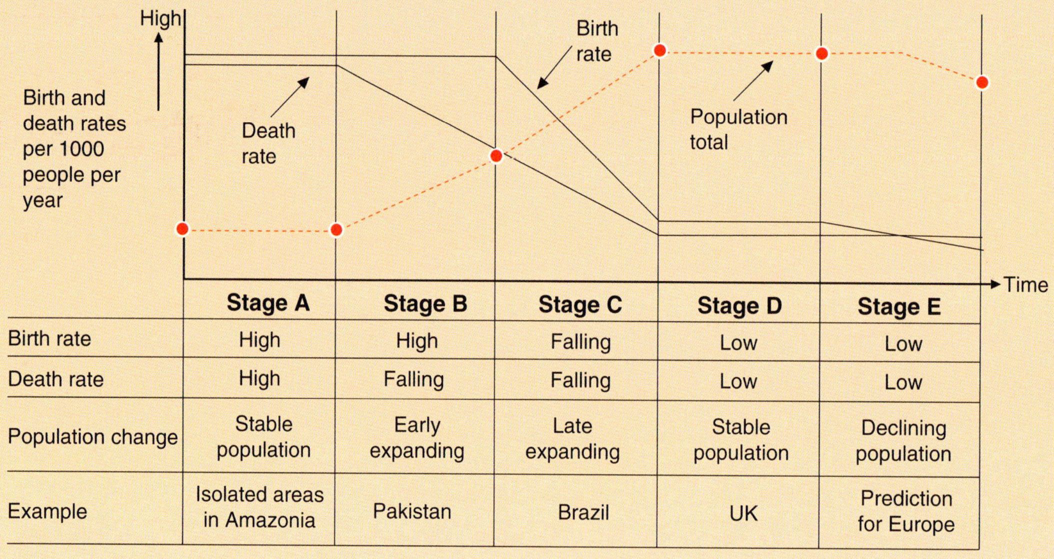

	Stage A	Stage B	Stage C	Stage D	Stage E
Birth rate	High	High	Falling	Low	Low
Death rate	High	Falling	Falling	Low	Low
Population change	Stable population	Early expanding	Late expanding	Stable population	Declining population
Example	Isolated areas in Amazonia	Pakistan	Brazil	UK	Prediction for Europe

This graph shows birth and death rates over a period of time. Most countries of Europe have moved through the Stages to Stage D. At the present time, some countries in Eastern Europe, such as Hungary, appear to have moved into Stage E - where there are fewer births than deaths. This may be a temporary result arising from the economic and social changes since the end of communism in those countries. However, if present trends continue, most countries in Western Europe might be expected to move into Stage E over the next few decades.

In societies remote from medical care and public health provision, such as in parts of Amazonia, there is a high death rate to match the high birth rate. This causes a stable population as in Stage A. When public health, nutrition and medical care improve, the death rate falls - as in Stage B. The birth rate remains high so there is rapid population growth. In Stage C, the birth rate starts to fall to match the death rate. The population is still rising, but not as rapidly. Once the birth rate has fallen to match the death rate, the population growth slows, as in Stage D.

Three countries compared

	Belgium	Kenya	Thailand
Capital city	Brussels	Nairobi	Bangkok
Population (1996)	10.2 million	28.2 million	60.7 million
Birth rate (per 1,000)	12	40	20
Death rate (per 1,000)	10	13	6
Annual rate of natural increase (%)	0.2	2.7	1.4
Infant mortality (per 1,000 live births)	8	59	36
Total **fertility rate** (av. no. of children per woman)	1.6	5.4 (1985 - 8.0)	2.2
Married women using contraception (%)	79	30	66
Time in years to double at present rate	630	25	48
Population structure:			
Percentage of population < 15	18	48	30
Percentage of population > 65	16	3	4
Life expectancy (years)	76	51	70
GNP per capita, 1994 (US$)	22,920	260	2,200
Official government view of birth rate	Satisfactory	Too high	Too high
Population density (per square km)	321	48	114
Adult **literacy rate** (%)	96	80 (m) 59 (f)	93
Urban population, 1994 (%)	97	27	20
Access to clean water (%)	100	28	77

Belgium

People: The country has two major ethnic groups, Flemish (55%, Dutch speakers) and Walloon (44%, French speakers). Belgium is an advanced, industrialised country with a high standard of living.

Health: Health services are highly developed with 1 doctor to every 310 inhabitants. An increasing proportion of the population are pensioners. The aging population puts strain on the health service because old people, on average, require more care than younger people. The ratio of population of working age to those too young or old to work is known as the **dependency ratio**. The working population has to pay an increasing proportion of taxation to pay for the health care, pensions and sheltered accommodation required for older people. At the same time, the aging population is less flexible and innovative in its work - so causing, perhaps, slower economic growth.

Kenya

People: Most Kenyans belong to the Bantu ethnic group. The most significant cultural and tribal groups are Kikuyu and Luo. 60% of the population are Christian. English is the official language although Swahili is the national language. In total, there are over 70 different languages spoken.

Health: Despite improvements, an estimated 30% of the population is malnourished. The birth rate remains one of the highest in the world. However, there has been a dramatic drop in mortality rates. Since 1963, the death rate has fallen from 22 people per thousand to 9 per thousand in 1993.

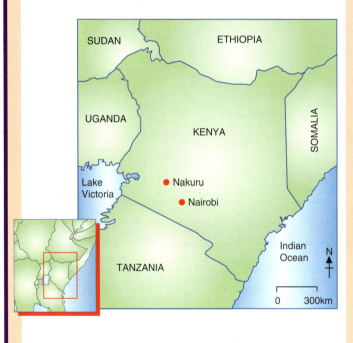

Thailand

People: Mostly ethnic Thai. Buddhism is the main religion.

Rapid urbanisation is taking place with an estimated 1 million living in slums or squatter settlements.

Health: The Thai Government is striving to achieve 'Health for All' by the year 2000. To this end, health services have been improved substantially over the past 20 years. All children are immunised against 6 major childhood diseases.

STAGE 5: What are the solutions?

Rapid population growth must be reduced or many more people will face famine and starvation. Around the world, governments have started to adopt policies for reducing population growth. India was the first country to declare an official population policy in 1952. Today, 129 countries have such policies.

One definition of a population policy is 'a deliberate effort by a national government to influence birth rate, death rate and migration'.

The resources that follow provide an outline of the population programmes for 2 countries, Kenya and China.

Briefly compare how these governments are tackling their population growth. Produce an information sheet that describes in more detail either China's or Kenya's population policy.

China's Population Policy

China started a 'One Child' policy in 1979 as a drastic means of lowering the birth rate. The plan was introduced in every factory, rural village and township, but has been most strongly enforced in urban areas. Government officials believed that, without the policy, it would take too long for the birth rate to fall and the population to stabilise.

The Policy

A couple first need permission to get married - they must wait their turn if local officials believe that people are getting married too young. The accepted ages are 28 for men and 25 for women. Once married, the couple must sign a 'one child certificate' agreeing to have just one child. The following resources are then available to them:

- priority for housing, education and health care

- a monthly payment of a subsidy

- higher pensions on retirement.

If the couple have signed the certificate but then have more than one child, the benefits are removed and financial penalties imposed. Most workplaces and communities have a 'quota' of allowed births and if this is exceeded there are financial penalties on them too.

After a woman has had one child, tremendous pressure is put on her to have an abortion if she becomes pregnant again. Some over zealous officials have been accused of forcing women to be sterilised after their first child is born.

One child policy

The policy has created large and controversial problems. One of the main ones is female infanticide because couples want their child to be male (and so carry on the family tradition). Scanners have been purchased in even the smallest communities to check the sex of unborn children. There is widespread evidence of women having abortions if they are having a girl.

If the policy continues up to 2025, the population growth rate will have declined to 0.5% per year, with an estimated population stabilising at around 1.5 billion. There will be more food and resources to go round, leading to a doubling of life expectancy and a rise in literacy rates. However, even with the policy in full force, China still expects to have difficulty in feeding its growing numbers.

One result of the Chinese population policy is that there are fewer girls than boys.

Problems for the future

There will be more men than women over the next decades. A shortage of brides might cause a 'marriage squeeze', threatening the country's moral and social fabric.

The shrinking family size may lead to difficulties as today's parents become older. There will be less support for them in their old age.

China's New Sorrow October 1996

Many feel that the 40% rise in the price of wheat this year has been caused by China's attempts to buy abroad what it has failed to grow at home.

Lester Brown of the Worldwatch Institute predicts that by 2030, China may need to import as much as 370 million tonnes of grain per year. This is equivalent to 80% of its present annual production. As the country's demand for grain increases, the outcome will be shortages and rapid price rises.

In response to this threat, the Chinese Government has set a production goal of 500 million tonnes by the year 2000.

Fertility rate for China (average number of children born to women)

Year	Rate	
1957	6.4	
1960	4.0	(fall due to famine and social upheaval in the Cultural Revolution)
1963	7.5	
1970	5.8	
1979	2.7	(start of the one child policy)
1980	2.5	
1994	1.9	

Population of China

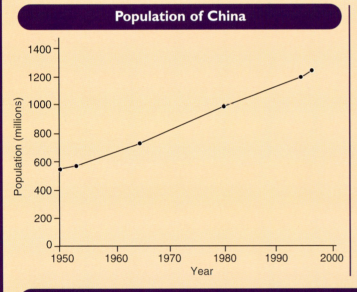

Kenya's population policy

Kenya's Family Planning Programme is one of the oldest in Sub-Saharan Africa. It started in 1965 but progress has been slow. Fertility rates have fallen from 8 to 6.5 children per family and the use of contraception has increased from 10% to 27%. Traditionally, women have improved their status in the community by bearing children. Projects aimed at population control try and find alternative means for women to achieve status, through education and employment schemes. The National Youth Service and local women's groups aim to reduce teenage pregnancy. This is a priority because one third of women still become mothers before the age of 18.

An example of an employment project to raise women's status and reduce high fertility rates

Kumuthanga is a small village 80 km south of Nairobi. In 1980, a women's group was given, by the charity Population Concern, a grant of £250 to start a business making and selling sisal baskets. The profits have partly been used to pay for the rainwater storage tank shown in the photograph. The women are very proud of their achievement. Their business has become a focus within the village providing status and security for women. Because they earn an income and are able to support themselves, there is less social pressure to have children. Fewer children are needed to provide labour on the farms or to support families as parents grow older. Teenage pregnancies are less frequent and the average family size has fallen from 8 to 5 children.

Two women stand in front of the rainwater storage tank.

Current policy goals of the Kenyan Government

- to reduce fertility levels
- to reduce internal migration which causes unplanned settlements in towns and cities
- to persuade and motivate males to use family planning
- to improve the status of women
- to extend the provision of education (particularly for women who currently receive, on average, only three years formal education on average)
- to widen the availability of contraception through family planning clinics.

However, a lack of political commitment and cultural barriers still threaten the success of these schemes. Having achieved dramatic falls in the death rate, Kenya's birth rate is still far too high, with the result that population growth is amongst the world's highest. The problem that Kenya faces is common to many countries. Without using the compulsory and, some might say, drastic methods adopted in China, Kenya can only use persuasion and the slow economic development of the population as means of reducing the birth rate.

STAGE 6: Review

At the start of this Enquiry, you were asked to take the role of a delegate to an international conference on population. Your task is now to bring together your work and prepare a presentation on the key questions relating to world population growth. These are:

- what are the rates of population growth in different continents?
- where in the world is population growing most rapidly?
- what are the causes of rapid population growth?
- how is population growth related to development?
- what are the possible solutions?

In your presentation, use maps, diagrams, posters, tables, graphs or any other means that you think are suitable.

Glossary

All the terms listed below are explained in this Enquiry. In each case, write a definition of the term and, if possible, give at least one example.

Birth rate

Death rate

Natural increase

Optimum population

North

South

More Economically Developed Countries (MEDCs)

Less Economically Developed Countries (LEDCs)

Population density

Demographic Transition Model

Fertility rate

Literacy rate

Life expectancy

Infant mortality

Population structure

Dependency ratio

Population studies-Migration

In the last Enquiry, birth and death rates were used as a measure of population growth. These are sometimes called 'crude measures' because they form the basis of population change. However, migration - the movement of people into and out of a country - also affects total population size.

The reasons why people leave an area are called *push factors*. The attractions of a new destination are called *pull factors*. These factors can vary enormously. They range from poverty and starvation, which trigger the flight from areas of famine, to the search for better incomes that fuels the movement of *economic migrants*.

During the nineteenth century and early part of the twentieth century, millions of Europeans migrated to the 'New World' of America. Many passed through New York where the Statue of Liberty symbolises the freedom and prosperity that people sought then and still seek today. You will see that the global pattern of migration changed in the second half of the twentieth century. In particular, Europe attracted many migrants. *Immigration controls*, designed to reduce the number of immigrants, were introduced in most developed countries.

Migration can occur within one country as well as between countries. This Enquiry concentrates on international migration. Internal migration within a country is covered in the Enquiry on Urbanisation (see pages 44-51).

questions to consider

1 Where have been the biggest migrations in recent times?
2 Why do people migrate?
3 What effect do migrations have on the source and the recipient countries (ie, the countries that people leave and the countries they move to)?

key ideas

People migrate because of *push* and *pull factors*. The push factors are those that persuade them to leave. Pull factors are those that attract them to a new place.

Most migrants can be classed as either *economic migrants* (people moving to improve their standard of living) or as *refugees* (people fleeing from persecution, war or famine).

activities

Using the information on this and the facing page:

Make a list of countries and regions from which people have moved and a list of those to which they have moved. Which of the movements can best be classed as having economic reasons and which are mainly movements of refugees?

Migration

Dateline: January 1994
Thousands flee war in Yugoslavia
An estimated 50,000 have left the former Yugoslavia as civil war erupts again in Bosnia. Most have gone to Germany.

Dateline: August 1995
Migrants enter Europe through southern Italy
Every week, thousands of Kurds, Albanians, Indians, Moroccans, Chinese and Yugoslavs pay $1,000 each to cross the Adriatic from Albania to Italy. This is the weakest link in 'Fortress Europe' where an estimated 9 million immigrants already live. As Europe strengthens its immigration controls, there is money to be made by criminal gangs who organise the illegal trade in people.

Dateline: December 1996
The long voyage
The first boatload of West Indian migrants landed at Tilbury in 1948. They were welcomed by newspapers and politicians as the solution to a labour shortage. Many worked in the National Health Service and in the transport system.
Now, their children and grandchildren are British but are often subjected to abuse and discrimination.

Dateline: December 1994
Australian economy a magnet for boat people
Now that Hong Kong is closed to them, Vietnamese migrants have started to travel to Australia. This has put pressure on Coastwatch, the government agency that prevents illegal entry, because Australia has thousands of miles of uninhabited coastline. The Vietnamese are able to land from boats and, if they manage to get to the major cities, they can settle and find work in the already established immigrant communities.

Global migration patterns

In recent years, international migration has become an important political topic. For example, in the UK, racial tension combined with high unemployment has caused governments to impose tight immigration controls. Many other countries have done the same and have set up systems to control illegal immigration.

In the nineteenth century and the twentieth century - up until about 1970 - tens of millions of people migrated. Many went from Europe to the 'New Worlds' of America and Australia. There was also massive movement within Europe, for example from Ireland to Britain, and from Mediterranean countries to northern Europe. More recently, many migrants came to the countries of Western Europe from the former colonies of India, Africa and the Caribbean.

This Enquiry looks at migration on a world scale and provides two detailed case studies on present mass migration.

The outcome of the Enquiry will be an article for a newspaper on one of the migration case studies.

STAGE 1: What are the patterns of global migration?

Present day migrants can broadly be divided into economic migrants and refugees. The economic migrants are those who seek to improve their standard of living by working in another country. Often, but not always, these migrants are accompanied by their families. Refugees are people who have fled from persecution, war or famine (sometimes all three at the same time!) In this type of migration, it is usual for the people from whole families, villages or towns to move.

Most of the richer, economically developed countries have experienced high levels of unemployment in the 1990s. This has led them to set up strict immigration controls, ie barriers to entry. These are aimed, partly, at avoiding the tensions that arise between unemployed people and the migrants who are sometimes accused of 'taking their jobs'.

One problem that arises in this topic is the difficulty in defining a migrant. Every day, millions of people travel across international boundaries. Many are holidaymakers, some are business people, some are temporary workers who will be abroad for a few weeks or months and others will be visiting relatives. At the same time, there will be people moving permanently to a new home and others, for example fleeing a civil war, who do not know when they will return home.

Because countries record and classify migration in so many different ways, global figures are always an estimate. On top of this difficulty, there is the problem of illegal immigrants. Nobody knows for certain how many there are - in the United States the number is probably several millions each year.

The resources which follow contain information on large scale migrations plus the reasons for them. Choose four migrations and briefly describe the reasons why people moved. In each case, suggest what the push and pull factors were.

Ethnic minorities in Britain

Most of the immigrant communities in Britain consist of the children and grandchildren of people who migrated in the 1950s and 1960s. Most came to Britain to improve their standard of living and many were directly recruited for jobs in textile mills, the National Health Service and the transport system.

They left their home countries to escape rural poverty, unemployment and poor health and education systems.

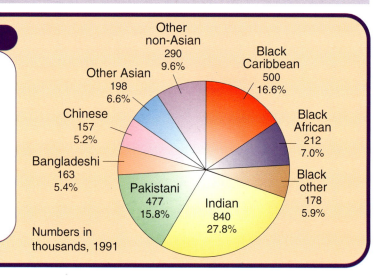

Other non-Asian 290 9.6%
Black Caribbean 500 16.6%
Other Asian 198 6.6%
Black African 212 7.0%
Chinese 157 5.2%
Black other 178 5.9%
Bangladeshi 163 5.4%
Pakistani 477 15.8%
Indian 840 27.8%
Numbers in thousands, 1991

Major migrations since 1950

Gastarbeiter in Germany

The **gastarbeiter** (guest workers) in Germany have largely been recruited from the Mediterranean countries, Italy, Yugoslavia and, the biggest numbers, Turkey.

In the early 1990s there were nearly 2 million Turkish people living in Germany. They were recruited to work in the car plants and factories of the booming German economy. At first, it was mainly young men who migrated, often planning to work only a few years and then return home with the money saved. The money they sent home was a great benefit to the Turkish economy. However, in more and more cases, they decided to stay and bring wives and family from Turkey to live with them.

When the German economy went into recession in the 1990s and unemployment rose, social and racial tensions increased. Often it was the Turkish workers who became the target of abuse.

These guest workers live in a hostel. They have left their families behind.

The problem was made worse after the collapse of communism and the reunification of Germany in 1989. Millions of East Germans and ethnic Germans who had been trapped in the former Soviet Union also moved to the West. Sometimes these people were in direct competition with Turkish people for their jobs. For these reasons, the German government has made it much harder for Mediterranean immigrants and, at the same time, has given financial incentives for Turks to return home.

Israel and Palestine

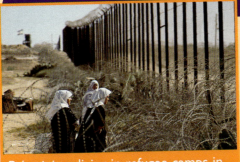

Palestinians living in refugee camps in Gaza.

In 1948, Jewish people succeeded in establishing the State of Israel. Since then, millions of Jews have returned to their 'homeland' from Europe, Russia, North Africa and the USA. In 1990 alone, more than 200,000 Russian Jews migrated to Israel. It had long been a dream of 'Zionists' to create a Jewish state where they would be free from persecution.

However, many of the residents of the region, also known as Palestine, fled in 1948 and set up refugee camps in Lebanon, Jordan and Egypt. An estimated 2.6 million Palestinians still live in these camps, though, of course, it is largely the children and grandchildren of the original inhabitants who are living there now.

In the mid-1990s, two of the biggest population movements were the flow of economic migrants from Mexico (and the rest of Latin America) to the USA and the flight of refugees from Rwanda to Zaire, Tanzania, Burundi and other neighbouring countries.

Below, you will find detailed accounts of these migrations.

Compare the two population movements in the form of a table. Set it out in the following way:

	Mexico to the USA	Rwanda to Zaire, Tanzania and Burundi
Number of migrants		
Reasons for migration		
Difficulties faced by migrants		
Impact on 'home' country		
Impact on 'host' country		

Jobless Mexicans defy the Border Patrol

Gabriel and Julian are typical of the thousands of Mexicans who try, every night, to slip across the border into Southern California. They are both 19, unemployed and from a farming area in Central Mexico. They cannot speak English and have little formal education but know, if they can get through the border, that there is work on the farms of the Central Valley, California, picking fruit, beans and cotton.

Sometimes, the would-be illegal migrants try to cross the border several times in one night. Occasionally they are lucky but, most often, they are caught and arrested by the border guards. The next morning they are then deported back across the border.

The US Border Patrol is the government agency whose job it is to stop illegal immigrants. Along parts of the US - Mexico border a 12 ft high metal fence is the first line of defence against the 'illegals'. The Border Patrol also uses searchlights, helicopters and tracker dogs. Nevertheless, an estimated 300,000 manage to get through every year.

Mexican migrants play a vital role in the economy of California, Texas and other border states. They carry out the jobs that Americans are reluctant to do, working in the hot fields or doing domestic work at wages below the official minimum. Even at these wages, the migrants earn ten times more than they can at home.

Because many of the migrants are young men, social problems arising out of drunkenness and boredom often arise. At the same time, their home towns and villages are losing those with the most initiative and enterprise, leaving behind an unbalanced population of the very young and old.

US fights to hold the line

Every night, the US border patrol holding camp at El Paso, Texas, is host to between 200 and 300 illegal migrants. Isidro Reyes is one such detainee. He is 28 years old and has left a wife and three children in the small mountain town of Madera, which is 200 miles to the south. He used to grow maize but last year there was no rain so he could not plant his seeds. In order to survive, he got a job in the construction industry but earned so little that, in his own words, 'You can either eat or buy clothes, but not both'. He knows that a day's pay in El Paso is the equivalent of two week's wages in Madera.

The promised land

For a very few, those clever or lucky enough to be successful, the rewards of migrating to the United States are beyond their wildest dreams. A big house with a swimming pool and a Cadillac are what some hope for. Others just want a steady job. But, for each success, fifty are left by the wayside and fifty return home defeated.

They did not know how much they would be abused by their employers and persecuted by authorities, how much effort it would take to reunite their families. They would lose contact with their old friends and traditions, they would be treated as the lowest of the low. Illegal migrants cannot complain if they are low paid, sacked unfairly or abused. There is always the threat that they will be sent back home.

Many illegal migrants in Los Angeles and other cities live in the 'barrio', a slum area in an industrial zone where the lucky ones are employed without anyone asking questions. At night it is dangerous to venture out, with gangs of bored youths roaming the streets, drunks and drug addicts. The air is full of the sounds of people arguing, women screaming, children crying, sometimes gunshots and police sirens.

Adapted from Isabel Allende, **The Infinite Plan**.

The US border patrol in Southern California.

Rwanda exodus overwhelms the United Nations

In 1994, one of the largest and most sudden migrations ever seen took place out of Rwanda. Social and ethnic violence, followed by civil war caused over 2.4 million people to flee from their homes. In one 24 hour period, more than 250,000 people crossed the border into Tanzania, Zaire and Burundi. The United Nations Secretary General declared that 'It is a humanitarian catastrophe'.

Relief agencies and neighbouring countries were completely overwhelmed by the flow of people. At the border crossing points there was chaos with 50,000 people waiting to cross one bridge into Tanzania. Food, shelter and emergency medical help could not be transported quickly enough to cope with the numbers. Many died from disease and starvation before the relief effort grew.

Most of the refugees were part of the Hutu social and ethnic group which, since 1990, had carried out violent and brutal massacres of its Tutsi rivals. The Tutsis, for their part, had formed a rebel army to protect themselves and drive the Hutus from power. As the rebels advanced, the Hutus - even those who had not taken part in the massacres - fled, fearing revenge attacks.

The origins of the conflict are long and complex. It is not simply one of rival **ethnic groups** fighting for power. Nevertheless, the effect has been to divide the country and make reconciliation almost impossible to achieve.

By late 1996, the patience and resources of neighbouring countries was exhausted. This time, most of the refugees were driven back into Rwanda where they faced an uncertain future.

Goma - a refugee camp in Zaire

By late 1994, an estimated 1.2 million refugees were living in Goma. Using shelters made from canvas, branches and leaves they relied totally on food brought in by the United Nations and other relief agencies. Apart from famine, the greatest threat was from water-borne diseases such as cholera and typhoid. Almost no drinkable water was available because it was contaminated with sewage.

Zaire is a very poor country and was completely unable to feed and house the refugees. Local people resented the newcomers because the camps covered their farmland, and also because the refugees destroyed the forests in their search for wood to use as fuel.

In Rwanda, the economy was crippled by the sudden loss of population. Whole villages emptied and fields were left uncultivated.

A refugee camp.

The sudden migration of so many people overwhelmed the neighbouring countries.

STAGE 3: Review

As a way of summarising the issue of international migration, you are asked to write an article for a weekend supplement of a national newspaper. The article should be about 400 words in length and should describe a major population movement. You need to include information on the numbers involved, the reasons for the movement and the impact on both the source and recipient countries.

Glossary

All the terms listed below are explained in this Enquiry. In each case, write a definition of the term and, if possible, give at least one example.

Push and pull factors in migration

Economic migrant

Refugee

Immigration controls

Gastarbeiter

Ethnic minorities

Ethnic group

Coursework Enquiry

Hypotheses: Most migrants move only short distances.
Most migrants move when they are in their twenties.
Most migrants move for economic reasons.

For the purpose of this study you must explain what a migrant is. One definition could be that a migrant is anyone who has moved home, whether internally within the country or as an immigrant from abroad. It is sensible to define a minimum distance for the move - say 20 kilometres - because moving within one town or city is not usually classed as migration. You have also to decide whether you will include temporary migrants and, if so, for how long the move must be.

The Enquiry can combine different research methods. You can collect primary data by using questionnaires and interviews. Before starting the research you should design and test a questionnaire to make sure that the questions are clear and it provides you with the data that you need. You can also use secondary sources for your data and then compare your results with the published material. You might find *Social Trends* and *Population Trends* in your local library.

Method of enquiry and report writing

1 Explain your hypotheses - why might the hypotheses be true? Make clear what definitions you are using and relate them to your region. Modify the hypotheses, if need be, to your local area. For instance, if you live in a coastal resort most of the migrants might be elderly. If you live in an industrial city, many of the migrants might have moved there many years ago as children. Will you ask about individuals or families? Will your research include the ethnic background of the people you survey?

2 Decide what data you need - how many people will you ask to complete the questionnaire. You need a good sized sample but do not want to be overwhelmed by information. The place of birth, the age at the time of migration, the distance moved, the number of moves and the reasons are all required. The 1991 Census will provide secondary data for your local area.

3 Data collection - prepare and test a pilot questionnaire before the proper survey. Make sure that you think about and plan the logistics, ie how many questionnaires to print, how to distribute and collect them, who to ask? Note that people sometimes are reluctant to give personal details so you must be sensitive and careful in the way that you approach them.

4 Collect the data - sharing the research with others will provide more data but needs careful organisation. You may need advice from a teacher or librarian to seek out secondary sources.

5 Data presentation - maps, graphs and tables can all be used. Try and use a variety of techniques such as flow line maps (to show the distance and number of migrants), scatter graphs (to show age and distance moved) and star / spider diagrams (to summarise reasons for moving).

6 Describe and analyse the data - you need to describe your findings in a written report and explain what they show. Do most people move due to 'push' or 'pull' factors?

7 Accept or reject the hypotheses - make a judgment on whether your data supports the hypotheses or not. It is also perfectly valid to say that there is not enough evidence to make a judgment and that further research is needed. Say which parts of the coursework were successful and which were not - and why.

8 Follow up - a small number of in-depth, follow up interviews will provide details of particularly interesting migrant case histories.

2 Urban environments

to the student

Most people in the UK live, work and shop in urban areas. Therefore the way that land is used in towns and cities affects nearly all of us. In this Enquiry, you will look at urban land use patterns. In other words, you will study where and why shops, offices, factories and housing are located in urban areas.

You will also be asked to consider issues such as the quality of life in different residential areas. What, for example, is the difference between living in an inner city area and a suburban housing estate?

The pattern of land use is dynamic, in other words, it is always changing. Conflicts of interest (ie disagreements) over how land is used often arise. You will be provided with information on one of the biggest issues - the building of shopping malls on the edge of towns and cities. Are they 'a good thing' and who do they affect?

questions to consider

1 What are the patterns of urban land use within the UK?
2 How does the pattern of urban land use compare in Bombay and Los Angeles?
3 What factors affect land use in urban areas?
4 What changes are occurring in urban land use in the UK?

key ideas

Urban area - a town or city. In the UK there is no agreed definition of what population size or density makes an urban area. All urban areas contain a mixture of land uses and there is sometimes conflict between these.

Planning controls - for the past 50 years, in the UK, local councils have had the power to control any changes in urban land use. Planning permission must be obtained before any new development takes place. People who disagree with a council decision can appeal to a government minister. Most urban areas in the UK grew up before planning controls were introduced.

Urban land use patterns - most towns and cities in the UK share some similarities in the way they have grown and in the way land is used. However, there are also differences due to their physical site and history.

activities

Using information from this and the facing page:

a) List the different ways in which land is used in urban areas.
b) With a partner, 'brainstorm' a list of possible land use conflicts in urban areas.
   ('Brainstorm' simply means thinking quickly 'off the top of your head'!)

URBAN LAND USE

URBAN LAND USE

How is land used in urban areas?

In this Enquiry you will investigate urban environments. You will study what they are and how they are changing. You will also be asked to consider reasons for the patterns of land use and for the changes. Most of the examples are from the UK, though in some cases it is useful to look at foreign cities. This is especially true where they can illustrate land use patterns before mass car ownership, for example in Bombay, India. At the other extreme, we shall also look at Los Angeles, a city that has a land use pattern influenced by far higher car ownership than that found in the UK.

The outcome of the Enquiry will be a report on the advantages and disadvantages of building a new shopping mall at the edge of an urban area.

STAGE 1: What are the patterns of land use in urban areas?

Most urban land use can be divided into:
- **residential**; **commercial** (shops and offices); **industrial** (factories and warehouses).
- open space (eg, parks); public buildings (eg, libraries, town hall); transport and communication (roads, railways).

Because the site, situation and history of urban areas are so varied, land use patterns are often complex. It is easier to start your study by looking at models of land use rather than particular towns and cities. A **model** is a simplification of reality, designed to help understanding. It does not include complex details so is a useful starting point.

In the resources that follow, you are given two urban land use models. You are also shown a simplified land use map for Preston in Lancashire. Write a brief description of the land use pattern in Preston. Then say whether the pattern you are describing is most like one of the models - or whether it combines elements from both.

Land use in Preston

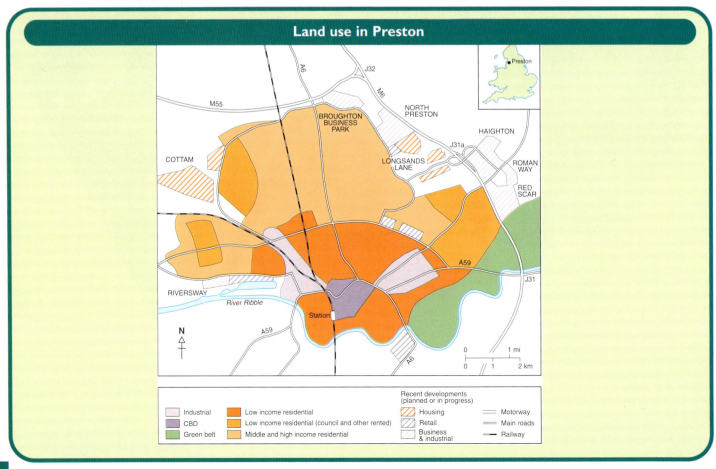

Site

The **site** of Preston is on fairly flat land above the River Ribble's flood plain. The town centre is approximately 30 metres above sea level.

Situation

Preston is **situated** in North West England between the Pennines and the Irish Sea. It lies on the north bank of the River Ribble at the lowest bridging point.

Urban land use models

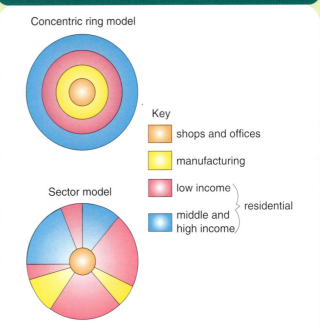

The diagrams show two urban land use models.

The first is called the Concentric Ring Model. At the centre is an area of shops and offices called the **Central Business District** (CBD). Around the CBD is an industrial zone of manufacturing with factories and warehouses. Lower income groups live nearer the centre, sometimes this housing is mixed in with the industrial zone. High and middle income groups live towards the edge of the urban area.

The second is called the Sector Model. At the centre there is a CBD but, instead of concentric circles, the zones of different land use are in the shape of segments. These stretch out from the CBD to the edge of the urban area. Often the segments are located along transport routes such as railway lines, canals or major roads. Notice that, just as in the Concentric Ring Model, the higher income residential areas are located away from the manufacturing zone.

What are the reasons for present day land use?

In towns and cities in the UK there are three main reasons why land is used in one particular way rather than another.

1 Land values
Anyone who wishes to use a particular plot of land must buy or rent it. The more that people want the plot of land, the more they will be prepared to pay for it. The busiest and easiest to reach locations are in the greatest demand so cost the most money. Shops are one of the most profitable uses of land. So, people who own shops can generally afford to pay more than people who want the land for housing, offices or factories. Therefore, the most accessible locations are usually shopping centres.

2 Planning laws
In the UK, new developments require planning permission. Councils often 'zone' areas for particular land use. So, even if someone could afford to build a factory in a busy shopping area, they would probably be refused planning permission.

3 Historical factors
Most towns and cities in the UK are very old - hundreds of years in some cases. Historical land use patterns might exist today simply because of decisions made long ago. Often, the oldest part of the urban area forms the present day CBD.

STAGE 2: Residential land use

Residential land use simply means an area of housing. The traditional pattern for over 100 years in the UK has been for high income groups to escape the factories, noise and traffic of inner cities to live in the suburbs. They have left behind lower income groups who could not afford to move. At the same time, new groups such as students and immigrants have moved in. These now live in the poorest and oldest housing nearest the Central Business Districts.

However, this pattern of high income groups living in suburbia and low income groups living in the inner cities has been affected by slum clearance schemes. In the 1950s and 1960s, many councils knocked down slums and rehoused the inhabitants in new council estates at the edge of towns and cities. Some of these have now become run down and suffer from the same problems as inner city areas. With high unemployment and low income levels, many of the council estate inhabitants are in a worse position than those left behind in the inner city. Because they rely mainly on expensive public transport and they are so far from the city centre, they are left trapped on the estates - away from shops and possible employment.

In the resources that follow, you are given information on the most and least deprived wards (electoral districts within a district council or borough) in Greater London. Use the resources to draw up a table that summarises the information. Use the following headings:

	Most deprived wards	Least deprived wards
Housing		
Residents		
Amenities		
Environment (eg open space / traffic)		

Then, briefly describe where the 100 most and least deprived wards are located.

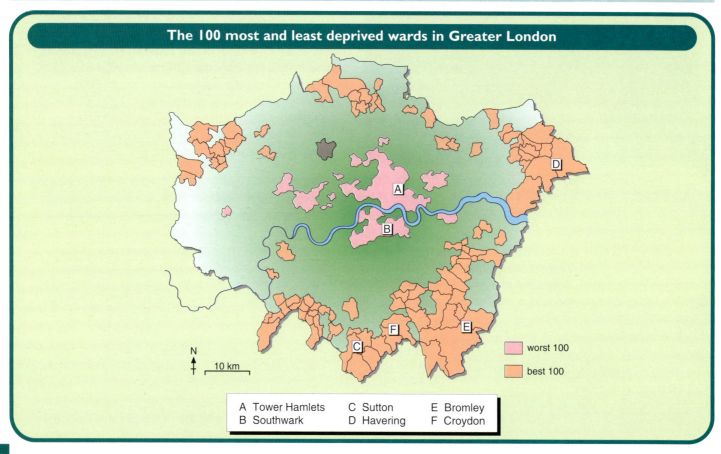

The 100 most and least deprived wards in Greater London

N
10 km

worst 100
best 100

A Tower Hamlets   C Sutton   E Bromley
B Southwark   D Havering   F Croydon

Ranking of Greater London wards in terms of deprivation: the lowest ranked five wards and the highest ranked five wards (there are 764 wards in the whole of Greater London).				
Ranked ward	**Borough**	**Unemployment (%)**	**Households not owning own home (%)**	**Households not owning a car (%)**
1 Spitalfields	Tower Hamlets	33	82	74
2 Liddle	Southwark	31	96	74
3 St Dunstan's	Tower Hamlets	27	84	69
4 Weavers	Tower Hamlets	25	82	67
5 St Mary's	Tower Hamlets	22	79	69
760 Woodcote	Sutton	6	5	5
761 Cranham	Havering	5	3	13
762 Biggin Hill	Bromley	5	8	8
763 Cheam South	Sutton	4	7	11
764 Selsdon	Croydon	4	5	11

Source: 1991 Census
Note: There are also a number of very prosperous areas to the south west of London. These are located in Surrey and Berkshire.

Report on poverty in London

In a sample survey, 3,000 residents in London were asked about their standard of living. This included questions on unemployment, household income and social issues such as litter and crime. It was found that people living in the most deprived areas suffer not only from poor housing and lack of household possessions, but also from a higher risk of road accidents, litter problems and lack of green space. They have more health problems than people living in less deprived areas, they are more at risk from violence or street crime and they suffer worse public transport. This is sometimes referred to as **multiple deprivation**.

Old terraced housing in Tower Hamlets.

Most of the people in the deprived wards live in rented housing whereas most people in the least deprived wards own their own homes. The poorest areas contain the oldest housing, much of it built in the nineteenth century. The streets are often laid out in a grid pattern of terraced houses or tenements with no front gardens - so allowing traffic to speed past people's front doors.

In the least deprived areas, most of the housing is newer with gardens. It is laid out in ways to slow passing traffic - with cul de sacs and crescents. Background and traffic noise is much lower. In these **suburban** areas there is usually better access to open space such as parks and golf courses. Often there are new out-of-town shopping centres and industrial estates where work is available.

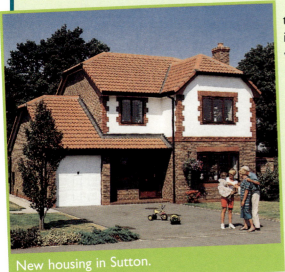
New housing in Sutton.

STAGE 3: Shopping patterns in urban areas

One of the most important **functions** (ie purposes) of urban areas is to provide shops and services. The traditional pattern in towns and cities is for a **hierarchy** of shopping centres to develop. A hierarchy is a rank order, in this case, of shopping centres. The smallest contains a single shop - for example, a corner grocer or newsagent. Next comes the suburban shopping parade with several shops, perhaps including a post office, off licence, dry cleaner and bank. At the other extreme there is the city centre shopping area covering many streets and containing hundreds of shops.

A corner shop sells everyday, cheap **convenience goods** such as bread and newspapers. These are called **low order** goods and are bought on a daily basis. People tend only to make a short journey to buy these items. The city centre shops sell more expensive, specialist goods such as clothes, cameras and furniture. These are called **high order goods** and are bought monthly or even less frequently. People are often prepared to travel several miles to buy these goods. Sometimes they are called **comparison goods** because people 'shop around' to compare range, price and value before they make a purchase. People do not generally do this for convenience goods such as newspapers, confectionery or basic groceries. National chain stores, such as Boots, and department stores, such as Debenhams, are usually located in city centres.

Since the 1960s, changes have occurred to the traditional shopping hierarchy. The development of supermarkets and edge of city shopping malls has made the pattern more complex. These changes, and the reasons for them, will be covered in more detail at a later Stage in this Enquiry.

In the resources that follow, you will find information about the location of shopping centres in south west London. Details are provided on the numbers and types of shops that are found in each location.

You are asked to present the information in graphs (pie charts or bar charts). Write a summary that describes the shopping hierarchy in this particular area. Explain why smaller shopping centres mainly provide convenience goods.

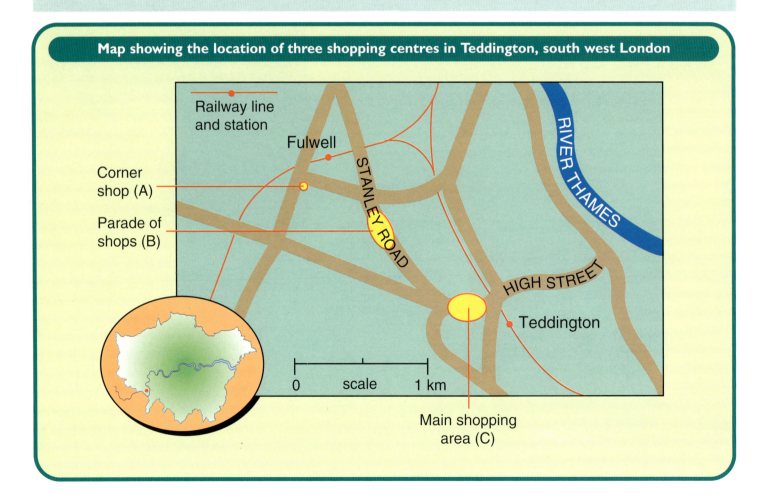

Map showing the location of three shopping centres in Teddington, south west London

Teddington - a suburb of south west London

Teddington is a fairly typical, middle income suburb of London. A large number of people who live there commute into London for their work. The area is mainly residential with some large open spaces. Within the Teddington area are a number of different types of shopping centre. These include:

- individual 'corner shops' selling newspapers and a range of groceries
- shopping parades with around 20 shops selling everyday, convenience goods
- Teddington High Street which is the main local shopping centre selling more high order goods.

In a survey of 3 shopping centres, the shops and services in 3 centres were counted and classified into those selling convenience and comparison goods. (A more detailed classification is provided in the suggested coursework enquiry on page 41.)

- The corner shop sells purely convenience, 'everyday' items.
- The shopping parade contains 23 shops, 15 of which sell convenience goods. There is one national chain store - the Thresher off-licence.
- Teddington High Street contains 41 shops and services. There is a much higher proportion of comparison goods shops - 24 in total. There are also a number of nationally recognised brand names on the High Street including Radio Rentals, Boots, Iceland, Tesco and Woolworth.

	Number of shops	Convenience	Comparison
Corner shop (A)	1	1	0
Shopping parade (B)	23	15	8
High Street (C)	41	17	24

The corner shop has a much lower **threshold** and **range** compared with, say, a shop selling electrical goods on the High Street. The threshold is the minimum number of people, living in the surrounding area, required to keep the shop viable. A corner shop needs about 500 potential customers to keep going. An electrical goods shop needs between 5,000 and 10,000 people. A large supermarket requires at least 50,000 people in its catchment area. The range is the distance that people are prepared to travel to buy something. Most people are not willing to travel far for a newspaper. But they are prepared to travel much further for more expensive items such as a television or washing machine.

Corner shop (A).

High Street (C).

Shopping parade (B).

High Street (C).

STAGE 4: How does the pattern of urban land use compare in Bombay and Los Angeles?

Bombay is a city of over 12 million people in India, (a less economically developed country). Vehicle ownership is 51 per 1,000 people (compared with 444 per 1,000 in the UK). Los Angeles is situated in southern California, one of the most prosperous areas of the USA. It has a population of over 11 million and a vehicle ownership of 783 per 1,000 people.

It is useful to contrast the two cities because they show the past and possible future land use patterns for UK urban areas. With its crowded streets, densely packed population and low car ownership, Bombay has a pattern similar to UK cities in the past. On the other hand, Los Angeles with its high car ownership, sprawling suburbs and freeway (motorway) system, shows what UK cities might become if present trends continue.

The two cities share some common features. Both are west coast ports and both are the home of their country's film industry.

The resources that follow contain maps and other information on Bombay and Los Angeles. You are asked to compare the two cities. In particular, compare the ways in which road transport has affected the pattern of land use. Summarise the information in the form of a short written article.

Los Angeles

The Los Angeles urban area stretches over 60 miles from north to south and over 50 miles from east to west. It contains over 50 malls, 10 of which compare in size with the biggest in Britain. The downtown area contains municipal buildings, offices and a relatively small number of shops. These shops tend to sell expensive, specialist goods for the office workers. The big suburban malls are far more important shopping centres than the downtown area. For example, the Beverly Center in Beverly Hills contains department stores and over 200 other shops.

The biggest malls in Los Angeles are mostly on the west and north side of the city where mainly middle and high income people live. People in the city rely almost totally on their cars for transport - most families own two cars. The urban land use pattern is dominated by cars and freeways. The city is criss-crossed with freeways and other major roads. As more and more people have moved to Los Angeles, new sprawling suburbs have been built out along freeways far into the surrounding desert.

The residential areas are **segregated** (ie, separated into zones) by income level and, to some extent, by ethnic grouping. The area to the south of downtown, including Watts, is where most low income, African Americans live. East Los Angeles is largely a low income Hispanic area.

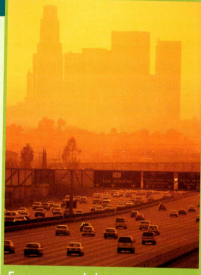
Freeways and downtown.

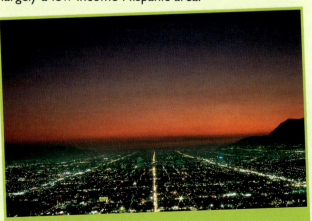
Urban sprawl.

Beverly Center.

Bombay

Bombay is built on seven islands. It is very compact compared with Los Angeles, even though it has a slightly larger population. It measures 20 miles north to south and 9 miles east to west. Half the population is homeless or lives in shacks. The biggest slum (or 'bustee'), Dharavi, houses 600,000 people in less than one square mile. The area is a mass of alley ways leading off roads which are packed with shops, street stalls and workshops. Unlike in Los Angeles, most people walk to work or travel by public transport. Their places of work are mixed in with their shops and homes.

   People who live in the suburbs mainly travel to work by bus or train. Every day, 4 million people commute by train and 3 million by bus. There are two main reasons for this. Firstly, average car ownership is low and, secondly, the roads are mainly narrow. They were designed for pedestrians and carts, so, although there are few cars compared with Los Angeles, the roads are often chaotic and jammed.

   Partly because the city is so tightly packed on its island site, rich and poor neighbourhoods, residential and industrial buildings, street markets and shops are all close together. The city does not have any urban motorways or suburban shopping malls like those in Los Angeles.

Shanty dwellings next to high rise apartments.

Congested traffic.

Maps of Los Angeles and Bombay

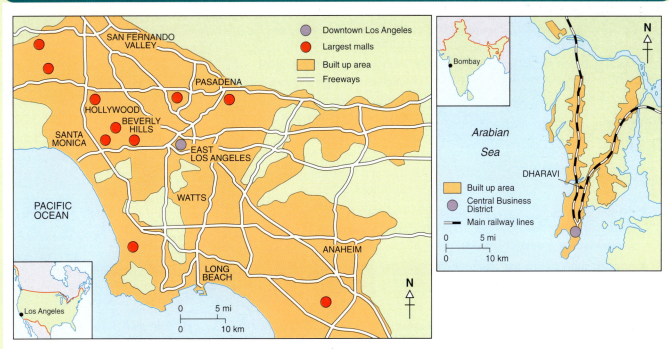

Note that the maps are drawn to the same scale. Although the populations of the two cities are roughly the same, Los Angeles is much bigger in area. This means that its population lives at a far lower density than that living in Bombay.

You have already seen the pattern of land use in a 'Third World' city, Bombay. That pattern of high density housing, mixed in with industrial and commercial land use, was typical of UK cities in the last century. Since then, the trend has been for the zones of land use to become segregated. In other words, the residential areas have become separated from the industrial and commercial areas. At the same time, the zones of high, middle and low value housing have also become more distinct as suburbs have spread out from the old urban areas.

If present trends continue, it is possible that UK urban areas will eventually become like Los Angeles. However, many people do not want this to happen. They say that the UK is a small country and we must protect our remaining green spaces from suburban sprawl. They wish to reverse the trend in car ownership and to improve public transport. Finally, they also hope to keep our city centres 'alive' with shops, entertainment and housing - rather than have the 'dead' CBD of Los Angeles.

Earlier in this Enquiry, you looked at land use in Preston, Lancashire. The resources that follow contain information on changes and developments in urban land use in Preston. These changes are typical of many towns and cities in the UK. Make a list of the developments and, in each case, briefly suggest reasons why the change has taken place.

If you live in or near an urban area, make a list of the recent developments that have taken place there and, again, suggest possible reasons for the developments.

Dockland redevelopment.

Reasons for changes in land use include:

- the desire by many people to live in green and pleasant suburban areas rather than in dirty and noisy inner city areas
- rising incomes allow more people to afford commuting (ie, travelling) costs on public transport
- rising car ownership gives middle and high income groups more freedom to live away from their work
- increased road building makes it easier for car owners to commute longer distances
- planning laws allow councils to control and plan development (allowing them, for example, to steer development into purely residential or industrial zones)
- council slum clearance schemes in inner city areas, followed by rehousing in council estates, have caused people to move.

Car ownership in the UK (% of households)

Preston Borough Council has recently published a Local Plan. It sets out the planning framework for new developments in Preston. Some of the points it makes are summarised below and opposite.

Green Belt - this is land on the Ribble flood plain and valley slopes which will be protected from urban development. The aim is to prevent sprawl to the east and therefore avoid merging with neighbouring built up areas.

Residential development - new housing will be needed in the future for the rising population. This must be concentrated in particular locations to avoid unrestricted urban sprawl.

Industrial and business development - this is concentrated in locations separate from residential areas. It has good access to motorways and major roads so avoiding delays and traffic nuisance for local people.

Dockland redevelopment - Preston Docks have closed for commercial traffic. This is mainly due to the use of larger ships which cannot navigate the narrow River Ribble. The old warehouses have been converted into housing or demolished to make space for residential and commercial developments.

Safeguarding the town centre - in the past, some out of town retail developments have been allowed. These have been located near major roads which allow easy access for motorists. However, the council now wishes to restrict such development and, instead, to promote the vitality of the town centre. In an effort to reduce congestion in the town centre, the council operates park and ride schemes. Pedestrian only zones, bus lanes and restricted parking are also used to reduce traffic in the central area.

Regenerating older residential areas - Preston has a large number of nineteenth century terraces which were built at the same time as the town's cotton mills. The mills have now largely closed. In the past, the Council demolished many of the older houses and rehoused people in estates on the edge of town. Today, the policy is - wherever possible - to renovate the old housing. This helps avoid urban sprawl and at the same time maintains a population near to the town centre. When run down, older housing is 'done up' (ie, improved) and higher income groups move in, the process is known as **gentrification**.

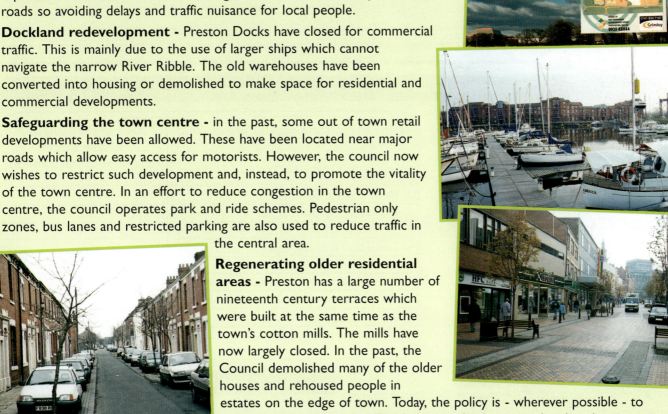

New land use developments in Preston

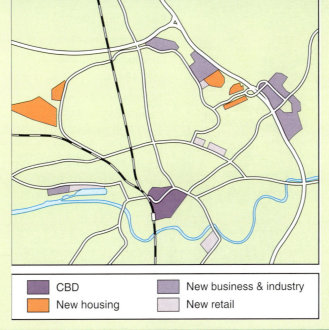

	CBD		New business & industry
	New housing		New retail

To help you bring together the issues contained in this topic, you are asked to write a report on a particular shopping development - Meadowhall outside Sheffield. Take the role of a planning officer who has been given the task of reviewing the development. Use the information provided. Set out the advantages and disadvantages of the development - who has gained and who has lost? Finally, state whether you would recommend similar developments in other areas.

Shopping malls in the UK (including those being built)

	Name of mall and nearest urban area	Area of shopping and leisure space (millions of square feet)
1	MetroCentre, Gateshead	1.6
2	Bluewater, Dartford, Kent	1.6
3	Merry Hill, Dudley, West Midlands	1.4
4	Lakeside, Thurrock, Essex	1.2
5	Meadowhall, Sheffield	1.1
6	Trafford Centre, Manchester	1.0
7	Cribbs Causeway, Bristol	0.7
8	Braehead, Glasgow	0.6
9	White Rose Centre, Leeds	0.6

Fact file on Meadowhall

Located east of Sheffield at Junction 34 of the M1.

Built in an area of high unemployment on the derelict site of Hadfield's steel works (closed in 1983).

Contains: 270 shops including Marks & Spencer, Debenhams, C&A, Boots and Sainsbury.

12,000 free car parking spaces.

Rail and bus station.

11 screen cinema.

Creche, parent and baby room.

18 restaurants / cafes / snack bars - spaces for 2,000 people. Open from 9 am to 8 pm (Mon - Thurs), to 9 pm (Fri), to 7 pm (Sat) and 11 am to 5 pm (Sun).

Your Personal Guide to Meadowhall.

Meadowhall cares for its customers

The growth of edge of city malls

Since MetroCentre opened in 1986, the pattern of shopping in Britain has changed radically. Out of town shopping now accounts for 25 per cent of all retail sales. Shoppers drive from 50 miles away to visit the malls, and buses bring people on special shopping trips from up to 200 miles away. In 1995, over 27 million visits were made to MetroCentre alone. Merry Hill attracted over 25 million visits and created 4,000 new jobs.

The top four shopping centres in Britain, in terms of profitability per square foot, are all edge of city malls. They are, in rank order, MetroCentre, Meadowhall, Lakeside and Merry Hill. By comparison, Oxford Street in the centre of London is only number 11 on the list. Meadowhall had, in 1995, a turnover of more than £600 million.

The shopping malls are starting to attract other development. In the USA, two thirds of all office development is next to edge of city malls. This trend has started in the UK with developments at Meadowhall, Lakeside and Bluewater.

Why is there opposition to malls?

In 1995, the Government announced that, in future, it would refuse planning permission for more edge of city shopping developments. This reversed a 10 year trend. The reason was that, in the view of many, the malls have started to 'drain the life' out of the traditional town and city centres. Research shows that since the opening of Merry Hill in 1989, Dudley has lost 70 per cent of its retail sales, Stourbridge 43 per cent, Birmingham and Wolverhampton 10 per cent each. So, while people have gained new jobs in the malls, other jobs have been lost on the high streets. If the trend continues, our city centre shopping areas will be boarded up except for a few charity and discount stores. They will be visited only by the elderly, the unemployed and those on low incomes - in other words, by people who do not have cars.

It is clear that, as more people own cars, the demand for new roads increases. This has the effect of creating ever more 'urban sprawl' - causing the loss of green countryside. Building more roads and edge of city malls encourages higher car ownership so the cycle continues. The country is being covered by motorways and other roads but the traffic jams are just as bad. So, it is said, the solution is to encourage more people to use public transport and to use the traditional high street shopping centres. By doing this there would be the additional bonus of cutting pollution from car exhausts and reducing traffic accidents.

The change in Government policy has been welcomed by the Association of District Councils. In the past they have often refused planning permission for edge of city developments only for their decision to be overturned by central government. Now they are more confident that they can halt the decline of town and city centres. Park and ride schemes, free parking and security cameras are all means by which local councils are trying to breathe life back into the traditional shopping centres.

Some views on shopping malls

It's nice and bright and its a safe place to bring children - shopper in Meadowhall.

It's just destroyed our trade in the town centre - shopkeeper in Dudley.

The malls are more fun - they remind me of the Fun Palace at the old pleasure beach - journalist.

The supermarkets and sheds are ugly and are taking away our countryside - Government minister, 1996.

The Bluewater development will create 6,800 jobs and attract 30 million customers a year - Government minister, 1995.

Millions of people shop in the malls every week because it suits the way they live - market research report.

Why are the malls so successful?

Market research shows that the malls meet the needs of growing numbers of people. Rising car ownership (especially of second cars) and the increasing number of working women make one stop shopping attractive. People can visit the mall and do their weekly shopping all in one visit.

Ease of parking and, more importantly, free parking persuade many people to visit the malls rather than the town or city centre. Additionally, the malls are 'weather free' where people do not have to worry whether it is raining or snowing.

Nowadays, more and more people view shopping as a leisure activity and the malls, with their cafes, cinemas and beautiful arcades have become places for 'a day out'. The changes to shopping laws which allow big shops to open on Sundays, together with their longer opening hours, have made this trend even stronger.

Many people feel more secure in malls than in city shopping centres. Security guards are often visible and this gives comfort to those shoppers who feel threatened by crime, drunks, beggars and vagrants on city streets.

The owners and developers of malls suggest that they are the pattern for the future. In earlier centuries, people shopped at market stalls. Then, in the nineteenth century, high streets of shops were built in the town and city centres. These were the most accessible locations for people on public transport. Now, in the age of mass car ownership and motorways, the mall is the most accessible and convenient form of shopping.

Glossary

All the terms listed below are explained in this Enquiry. In each case, write a definition of the term and, if possible, give at least one example.

Urban area	Suburban
Planning controls	Function (of an urban area)
Urban land use model	Shopping hierarchy
Residential land use	Low order / convenience goods
Commercial land use	High order / comparison goods
Industrial land use	Range
Central Business District	Threshold
Zoning	Segregation
Site	Green belt
Situation	Gentrification
Multiple deprivation	

Coursework Enquiry

Hypothesis: Larger shopping centres have a greater proportion of comparison goods shops than smaller shopping centres. (This, of course, also means that they have a smaller proportion of convenience or everyday goods shops.)

On page 32, there is a study describing 3 different sized shopping centres in south west London. These centres form a hierarchy. This is a relatively easy study to duplicate if you live in an urban area.

Background theory:

The reason for a shopping hierarchy is that people are not generally prepared to travel far for convenience goods (such as bread or newspapers) so shops selling these goods will tend to be local and scattered throughout the urban area. On the other hand, people are prepared to travel further for comparison goods (such as clothes or furniture) and they will also want to compare one shop with another. Therefore, these shops tend not to be in local areas but are found in bigger shopping centres.

Shops selling convenience goods (low order) include:

- grocers
- bakers
- chemists
- greengrocers
- butchers
- off-licences
- newsagents
- sub-post offices
- public houses

Shops selling comparison goods (high order) include:

- clothes / shoe stores
- jewellers
- specialist stores (eg, records and sports goods)
- electrical goods stores
- furniture stores

Method of enquiry and report writing

1 Explain your hypothesis - what are convenience and comparison goods? Why do you think the hypothesis will be true?

2 Decide what data you need - ie, total number of shops, total area of shops, number of shops of different types, classification of shops.

3 How will you collect the data? - from secondary sources (eg, published plans in a local library) or primary sources (eg, fieldwork observation). Do you need questionnaires or data collection sheets?

4 Collect the data - in groups or individually?

5 Present the data - in the form of maps, tables, graphs and transects.

6 Describe and analyse the data - what do your maps and graphs show?

7 Explain the data and evaluate your hypothesis - was the original idea correct? Is your hypothesis supported or rejected? Were there any weaknesses in the way that you collected and recorded data?

8 Collect further data if necessary - you might want to find out how often people shop in these centres, how far they travel and what transport they use.

③ Rural urban links

to the student

In Europe and North America living in cities is losing its attraction. Many people have moved out to the countryside or to smaller towns. As a result, cities in the more developed parts of the world are losing population. However, in the less developed parts of the world, people are still moving into cities in large numbers. Some of these cities are growing at a tremendous rate.

Whilst **London** and **Paris** were once amongst the largest cities in the world, today they cannot match the huge populations of cities like **Mexico City**, **Tokyo**, **Shanghai** or **Calcutta**. In the first Enquiry, we look at one of the fastest growing cities, Jakarta in Indonesia.

Although **90 per cent** of people in Britain still live in or near to urban areas, the larger cities such as **London** and **Birmingham** are losing population. We use Liverpool as an example in our second Enquiry.

questions to consider

1 Why do people migrate from rural areas to cities in less developed countries?
2 Why is there a movement of population and industry away from cities in more developed countries?
3 How do rural - urban migrations affect cities and the countryside?

key ideas

Urbanisation - this is a process in which an increasing proportion of the population lives in urban areas. It is caused by the growth of cities and the movement of people into cities.

Counter-urbanisation - this is a process occurring in many high income countries. People are leaving inner city and older urban areas and are moving to the suburbs and surrounding towns and villages.

activities

Using information from this and the facing page:

Working with a partner:

a) List some of the reasons why people might migrate to cities in less economically developed countries (LEDCs).
b) List some of the reasons why people may be moving out from cities in more economically developed countries (MEDCs).

Plantations replace peasant farms

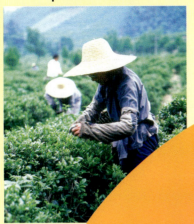

Drought and natural disasters

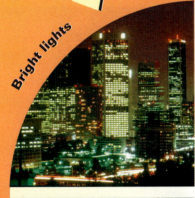

Refugees from war

People migrate from rural areas to cities in less developed countries

Bright lights

Jobs

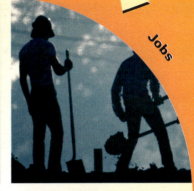

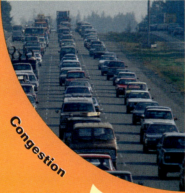

Congestion

Industrial decline

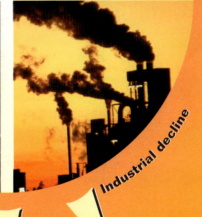

People migrate from cities to suburban and semi rural areas in more developed countries

Industrial estates and science parks

New towns

Leafy suburbs

Urbanisation - Jakarta, Indonesia

In this Enquiry you will investigate the world pattern of urbanisation. You will be asked to prepare a presentation that explains why cities are growing most rapidly in the less economically developed countries (LEDCs). In addition, you will explain how rapid urbanisation affects the people and local government of Jakarta, Indonesia and what people are doing to try and solve the problems that arise.

The outcome of the Enquiry will be a presentation in four parts:
1 Where are the world's largest and fastest growing cities?
2 Why is Jakarta growing so quickly?
3 What problems are caused by rapid population growth in Jakarta?
4 How have ordinary people and organisations tried to solve these problems?

STAGE 1: The world pattern

In 1950 there were 80 cities in the world with populations over 1 million. By 1995, this number had risen to 287. The following map and table show where the world's largest cities are located and which ones are growing most rapidly. You are also given some reasons why people move from the countryside (**rural areas**) to the cities (**urban areas**).

  You are asked to make some notes on the reasons why people move into cities. Which areas of the world were most urbanised in 1995? Briefly describe where urbanisation is most rapid.

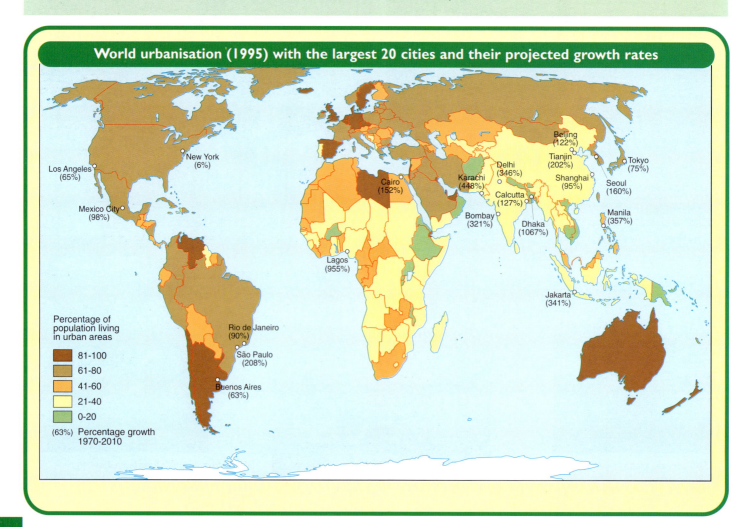

World urbanisation (1995) with the largest 20 cities and their projected growth rates

Beijing (122%)
Tianjin (202%)
Tokyo (75%)
New York (6%)
Los Angeles (65%)
Delhi (346%)
Karachi (448%)
Shanghai (95%)
Seoul (160%)
Cairo (152%)
Calcutta (127%)
Mexico City (98%)
Bombay (321%)
Dhaka (1067%)
Manila (357%)
Lagos (955%)
Jakarta (341%)
Rio de Janeiro (90%)
São Paulo (208%)
Buenos Aires (63%)

Percentage of population living in urban areas
- 81-100
- 61-80
- 41-60
- 21-40
- 0-20

(63%) Percentage growth 1970-2010

Between 1950 and 1970 New York was the world's largest city.

The growth of 4 major cities

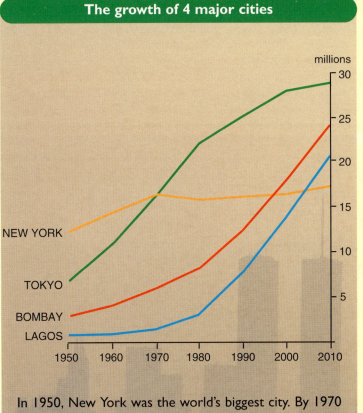

In 1950, New York was the world's biggest city. By 1970 it had been overtaken by Tokyo. Now, cities like Bombay and Lagos have growth rates much faster than Tokyo. If the trends shown on the graph continue, when will Bombay and Lagos overtake Tokyo?

Urbanisation by region 1950 - 1995 (with a forecast for 2025)

(Figures show the % of the total population who live in urban areas)

	1950	1995	2025
Africa	14	35	54
Asia	16	34	54
Europe	56	75	84
North America	64	76	85
South America	43	78	87
Central America	40	68	80
Caribbean	35	62	74
Former USSR	41	68	80
Australasia	61	71	77
World	**29**	**45**	**61**

Moving into cities - urbanisation

The large cities of Europe and North America grew rapidly in the nineteenth century and the early part of the twentieth century. Since 1960, the growth of these cities has slowed and, in some cases, has reversed as people leave to live in the countryside.

In the developing world, however, cities are growing at an alarming rate. Cities like Calcutta, Shanghai and Mexico City are much bigger than any known before in the history of the world.

Why do people move into cities? In the less developed countries, cities offer the chance of jobs, higher wages and better living conditions. Industries, administration and services tend to concentrate in the largest cities and attract even more people. Unfortunately many people cannot find jobs when they arrive. At least 20% of urban populations have no regular job and few governments pay any unemployment benefit.

But, even though conditions can be bad, city dwellers are usually better off than those who live in rural areas. When they can get a job, wages are usually much higher than in the countryside and there is a better chance of getting access to medical care and education.

And of course, life in the countryside is hard. As the population rises there is less land available and most good farmland is owned by a few rich landlords, sometimes multinational companies. There are few chances to get on and little chance of improving life. Sometimes natural disasters such as drought or famine in Africa's Sahel or flooding, as in Bangladesh, make people homeless. Civil wars in countries like Cambodia or the Sudan make life in the countryside too dangerous and refugees pour into city areas.

So a large number of factors act as a magnet or pull to the city whilst poverty and danger may push people from the countryside. Geographers refer to these as the **push and pull factors** causing **rural-urban migration**.

In Stage 1, you saw why people are moving into cities around the world. In the resources that follow you will find some specific reasons why people are moving into Jakarta, one of the world's fastest growing cities.

Use the information in the resources to make notes on Jakarta's growth with such headings as 'The pull of the city' or 'Leaving the countryside'. Some of the points could be made most effectively in diagram form or on an outline map of Indonesia. A graph to show the growth of Jakarta's population would also be helpful.

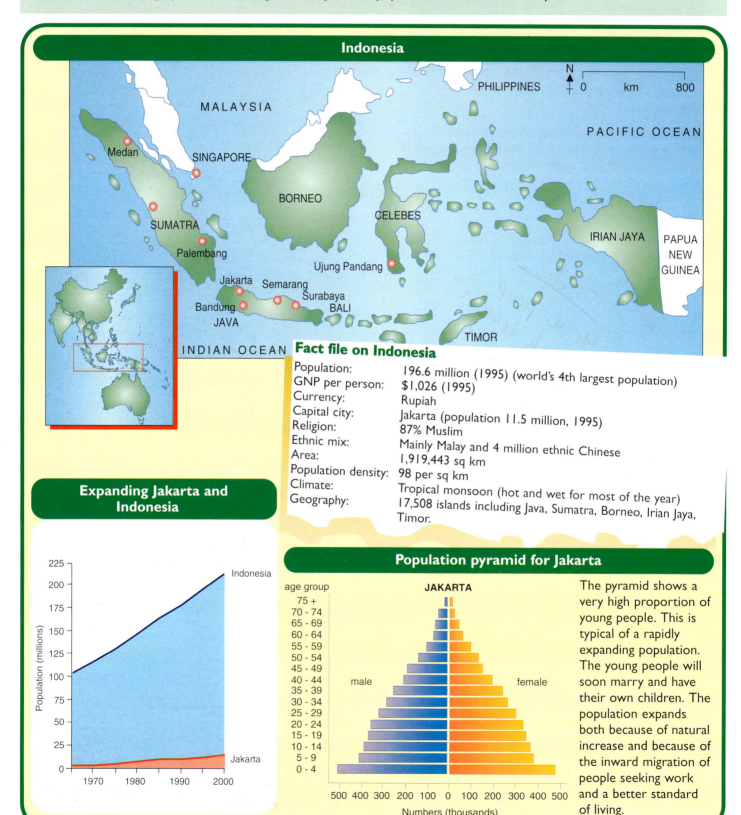

Indonesia

Fact file on Indonesia

Population:	196.6 million (1995) (world's 4th largest population)
GNP per person:	$1,026 (1995)
Currency:	Rupiah
Capital city:	Jakarta (population 11.5 million, 1995)
Religion:	87% Muslim
Ethnic mix:	Mainly Malay and 4 million ethnic Chinese
Area:	1,919,443 sq km
Population density:	98 per sq km
Climate:	Tropical monsoon (hot and wet for most of the year)
Geography:	17,508 islands including Java, Sumatra, Borneo, Irian Jaya, Timor.

Expanding Jakarta and Indonesia

Population pyramid for Jakarta

The pyramid shows a very high proportion of young people. This is typical of a rapidly expanding population. The young people will soon marry and have their own children. The population expands both because of natural increase and because of the inward migration of people seeking work and a better standard of living.

Rural areas in Java

Nearly all the work of planting, harvesting and maintaining the embankments is done by hand. This is extremely hard work in the tropical heat. Although the hot, wet climate is ideal for rice growing, and the rich volcanic soils are very fertile, most people in the rural villages have a low standard of living. This is because there are too many people for the land to support. As the population rises, there is not enough land to provide everyone with a plot big enough to support a family.

No wonder that many people are deciding to leave the village and search for new jobs in the large cities. Even so, 77% of the people in Indonesia still live in the countryside. Some of the landless have moved into the upland rain forests, clearing trees to create new farmland. Often this only results in soil erosion on the steep slopes, rapid run-off of rainwater into the rivers and flooding in the lowland areas the people have left. Indonesia is a country where poverty is being reduced by rapid industrialisation but 27 million people still live in **absolute poverty**, ie their lives are threatened by lack of food, housing and medical care.

Villagers know that most people living in the cities have easier lives than those who work in the fields all day. Therefore there is a strong temptation to leave the countryside and try for a better life in Jakarta or Java's other major cities.

Come to Jakarta

Jakarta's official population in 1995 was 11.2 million people but unofficial estimates put it at nearer 18 million. The extra millions are the homeless or those living in unofficial shanty town housing around the city's edges.

So why are people flooding into Jakarta from all over Indonesia, but especially from the rural areas of Java where 60% of Indonesians live? Most Indonesians have access to TV and the messages they get about Jakarta are these:

Modern homes - high rise apartment blocks are springing up all over the city. In both eastern and western suburbs there are new housing areas which can accommodate 75,000 people each.

Social facilities - most parts of Jakarta have access to modern schools, hospitals and clinics.

Recreation - modern golf courses, huge new shopping malls and modern night life with westernised images - the Hard Rock Cafe, Kentucky Fried Chicken, even Marks and Spencer, and the Indonesian version of the night-club - the Dangbut Bars with cheap beer and popular music.

Jobs - many multinational companies have factories around Jakarta with a particular emphasis on sports shoes (eg Nike and Reebok), clothing and consumer goods. Indonesia's own industries are growing too. The country has just completed the building of its first airliner.

Money - over a million people in Jakarta now form a middle class with money to spend. These people have an average disposable income of £170 per month. Whilst Indonesia's average income per person (GNP per head) is £430 per year, in Jakarta the average is £1,400.

STAGE 3: Living in Jakarta

We have seen that Jakarta has a population of over 10 million and this is rising rapidly. However, once people move to the city, they have many problems to face. On the outskirts of Jakarta is the shanty town suburb of Serang. Most of the people in Serang have arrived from the countryside during the past 10 years. Many of them are unemployed and try to make a living in the informal sector of the economy, eg by begging, selling food on the streets, scavenging from the refuse tips on the edge of the city or working as shoeshine boys or car windscreen cleaners at traffic lights.

In the resources that follow, many of these problems are outlined. Make two lists of the problems - one for the people themselves and one for the city government.

Eni

Eni is 17. She works in Jakarta's Eltin factory which employs 13,000 people making trainers for Reebok, Adidas and Nike. Indonesia is the world's leading producer of sports shoes.

Eni wears plastic sandals. She cannot possibly afford to buy a pair of the trainers she makes. Her wages are only 16p an hour and the trainers sell for between £50 and £100. The labour costs for a pair of these trainers is just over £1.

Conditions in the factory are overcrowded, hot and noisy. Temperatures are over 30°C. If Eni makes a mistake she is shouted at and even hit by the supervisors. She lives in the factory compound. Eni shares a bunk room with 18 other girls who, like her, have come to Jakarta from rural villages in order to find work.

Sadisah

Sadisah is 37. She came to Serang with her husband Rahman four years ago when they lost their farm on Sumatra because they could no longer afford to pay the rent. Sadisah has six children. They live in a two room hut that is made from plywood and tin. There is no electricity or running water and no sanitation. The family use a bucket for a toilet and this is emptied into the communal cess pit. At times of heavy rain the cess pit floods and the mud tracks around the **shanty** are awash with stinking sewage.

Sadisah's husband Rahman found a job as a street cleaner when they first arrived in Serang. But last year Rahman became very ill with a chest complaint. He developed asthma and what they think is

Sadisah and her family live in this shanty area.

bronchitis. Rahman blames his job for this. The streets are overcrowded with traffic and the air quality is very poor. Photochemical smog can often be seen. This is caused by the mixture of vehicle exhaust fumes and smoke from industrial chimneys. The family cannot afford medical care for Rahman who, at the age of 38 is unable to work.

Taska

Taska is Sadisah's youngest son. He is 9 and has a place at the local primary school like most other Indonesian children of his age. But Taska often misses school to go begging on the busy streets of Jakarta so that the family can afford enough food. Like many of his friends, Taska finds that jangling tambourines in car windows when they stop at traffic lights often earns a few rupiahs.

Iqbal

Iqbal is the only member of the family with a permanent job in the **formal** sector of the economy. That is a job that pays a wage, like Eni's in the shoe factory. He works on one of the city's 100 faeces trucks. There is no sewerage system in most of Jakarta. All human waste collects in cess-pits and Iqbal's job is to empty these and take the waste to one of the city faeces disposal dumps. He is paid the minimum wage of £1.50 a day.

Because the city does not have a sewerage system, cess pits are used.

Paina

Paina is Sadisah's oldest daughter. Often they do not see her for days and, although the family do not talk about it, they know that Paina works as one of the city's estimated 18,000 prostitutes. Many girls and boys as young as 10 or 11 sell sex in the streets of Jakarta. The city authorities have tried to regulate this by providing hostels for them, but most still live on the streets. These young people live dangerous lives in a city where crime, disease and drugs are widespread.

Marsinah

Marsinah also lives in Serang. She came here to escape fighting in one of Indonesia's mountainous islands where opponents of the government have been waging a civil war. Her husband was killed in the fighting. Now, Marsinah makes a living by making and selling 'bakso' - steamed noodles - on the streets of Jakarta.

Marsinah's shack is close to a well which a charity organisation helped the community to build. She uses the water from this well to steam her noodles. Unfortunately the water from the well is often contaminated with brine. Because so many wells have been sunk into the water bearing rocks, or aquifers, below Jakarta the water table has dropped and sea water often infiltrates the rock.

Eni and Iqbal have permanent jobs which pay a wage. These are jobs in the **formal** sector of the economy. Many of Jakarta's residents are lucky enough to have better paying jobs in the city's offices, shops, factories, construction and transport industries.

Marsinah, Paini and Taska manage to earn some money for themselves even though they do not have an officially recognised job. They work in the **informal** sector of the economy. Other informal jobs include scavenging on the city's refuse dumps for waste material to recycle, selling local crafts such as batik cloth to tourists and begging.

Eni, Sadisah and the rest can see, across the city from Serang, the modern apartment blocks built by private developers to house the city's workers. These apartments have electricity and running water. But to get one of these, they would need much better paid jobs and for that they would need education and training. This massive contrast between rich and poor is typical of big cities in less developed countries.

STAGE 4: Can the problems be solved?

In Jakarta, people are aware of the problems facing the poor, especially the condition of the housing and the lack of jobs. Action to try and overcome these problems is being taken at different levels. Individuals and groups of local people are trying to help themselves. The city authorities and national government are investing money, for example into housing schemes.

At the same time, work is going on in the countryside to solve the problems there. The rural-urban migration of young and enterprising people often means that only the old and weak will be left behind. Food production will fall and the people left behind will be worse off than ever. So, schemes to raise the rural standard of living and slow the migration are seen to be increasingly important.

In the following resources, you will find examples of how people might do something to help. Complete a table like the one below to summarise the points:

Example	Description of what is being done	Advantages	Problems

GETTING TO GRIPS WITH JAKARTA'S PROBLEMS

Most of Jakarta's population lives in modern housing.

1 The Jakarta Plan
The local government of Jakarta has drawn up a Master Plan for the city. Its priorities are:
- to solve the problem of slum areas
- to provide more jobs
- to improve public transport.

2 New homes for people
High rise apartment blocks are being built by private developers all over Jakarta. These apartments are relatively expensive to buy or rent and can only be afforded by the better off middle class families. There are over 1 million people in this group.

3 Public housing
Most squatter settlements - or shanty towns are illegal. The land is not owned by the squatters. The government has tried to deal with this problem in several ways:
- sending in bulldozers and riot police to clear the areas. BUT the people need somewhere to live and will set up new squatter settlements elsewhere, or move back the next day.
- providing cheaply built public housing, usually high rise, high density blocks with basic amenities. People are then persuaded to move BUT rent has to be paid for these flats which the poorest cannot afford, so often they move back to shanties as their debts increase.

4 Self help schemes
The people living in shanties are well established and do not want to move but they cannot improve their homes because they are afraid that any day the police might move them out.

If the government gives these people security of tenure - that is, the right to stay on the land - and help to provide building skills and materials, the people will improve their own housing.

JAKARTA - A SUMMARY

Progress
* Since 1967, per capita income has risen from $50 to $650.
* Absolute poverty has been reduced from 60% of the population to 17%.
* Life expectancy has risen by 20 years and the birth rate has fallen from 27.7 to 21.5 per thousand.
* Infant mortality has decreased from 33 to 31.8 per thousand births.
* Most 7 - 12 year olds now receive primary education.

Remaining problems
* 27 million people in Indonesia still live in absolute poverty - many will be tempted to move to Jakarta.
* Hundreds of thousands of people still live in shanty towns around Jakarta and other major cities.
* Jakarta does not have a modern sewerage system and water-borne diseases are common.
* The only safe drinking water is bottled - which most people cannot afford.
* Petty crime, child prostitution and begging are common.

Raising the standard of living in rural areas

Self help schemes in Indonesia, backed by international organisations like Traidcraft, based in the UK, are trying to raise rural incomes. This has the dual effect of raising living standards and of slowing the migration to cities like Jakarta. One scheme is run by Pekerti, a marketing organisation for crafts such as hand woven fabrics. Villagers are given technical help and training in management skills. The role of Traidcraft is to advise on products to make and then market them through its UK catalogue.

Such schemes are successful in raising villagers' incomes but sometimes suffer from a lack of funding and investment. Also, the market for the craft goods is very crowded with similar schemes running throughout Asia, Africa and Latin America. Nevertheless they do some good and, in most people's view, it is better to trade fairly rather than simply give charity. In this way, the recipients of help gain self respect and, if they are lucky, a starting point for future growth.

Villagers in central Java are given advice on designing and producing batik fabrics.

Housing in Jakarta

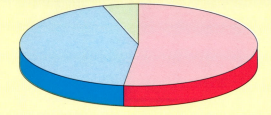

- Permanent (eg high rise apartments)
- Semi permanent (eg breeze block)
- Temporary (shanties)

STAGE 5: Review

So far, you have made background notes and gathered information on urbanisation, with particular reference to Jakarta.

Now, prepare a presentation in four parts:
1 Where are the world's largest and fastest growing cities?
2 Why is Jakarta growing so quickly?
3 What problems are caused by rapid population growth?
4 How have ordinary people and organisations tried to solve these problems?

The presentation can be made in several ways, for example: a talk, supported by maps, pictures and diagrams; a wall display with text, maps and graphs; an audio cassette with supporting illustrations; a written report.

Glossary

All the terms listed below are explained in this Enquiry. In each case, write a definition of the term and, if possible, give at least one example.

Urbanisation	**Rural-urban migration**
Rural area	**Absolute poverty**
Urban area	**Shanty**
Pull factors in migration	**Formal sector of the economy**
Push factors in migration	**Informal sector of the economy**

enquiry

Counter-urbanisation: moving out of Liverpool

Unlike Jakarta in Indonesia, the population of Liverpool in the UK is going down. This is true of most large cities in the UK, Europe and the USA. This Enquiry looks at why this is happening and how it affects cities like Liverpool. The outcome of the Enquiry will be a short essay with the title: 'Why are people leaving Liverpool?'

STAGE 1: The UK pattern

The city of Liverpool is the main built up area in Merseyside and is one of the UK's largest cities. Its population is declining, like most other cities. Using the resources which follow, describe and explain the main population trends in the UK since 1951.

Write a list of the major towns and cities in rank order by size of population in 1994. On an outline map of the UK, mark the location of the UK's largest cities. Mark those which gained population (between 1951 and 1994) with a black dot and those that lost population with a red dot. Briefly describe what your map shows.

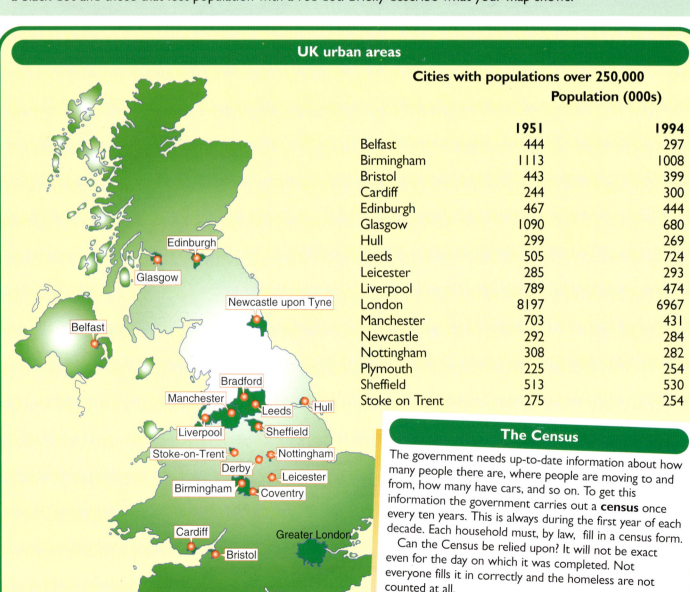

UK urban areas

Cities with populations over 250,000

	Population (000s)	
	1951	1994
Belfast	444	297
Birmingham	1113	1008
Bristol	443	399
Cardiff	244	300
Edinburgh	467	444
Glasgow	1090	680
Hull	299	269
Leeds	505	724
Leicester	285	293
Liverpool	789	474
London	8197	6967
Manchester	703	431
Newcastle	292	284
Nottingham	308	282
Plymouth	225	254
Sheffield	513	530
Stoke on Trent	275	254

The Census

The government needs up-to-date information about how many people there are, where people are moving to and from, how many have cars, and so on. To get this information the government carries out a **census** once every ten years. This is always during the first year of each decade. Each household must, by law, fill in a census form.

Can the Census be relied upon? It will not be exact even for the day on which it was completed. Not everyone fills it in correctly and the homeless are not counted at all.

Population movement in the United Kingdom

The North-South Drift
Over the last few decades there has been a southward drift of population. Many people have left the old industrial regions such as North East England and central Scotland. Many left because there was a shortage of jobs whilst others retired to the warmer and more rural areas of the south.

The urban-rural shift
The largest internal migration in the United Kingdom since the 1950s has been the rapid movement of people out of the large cities. Since 1961, London alone has lost one million people whilst the other major cities have lost 500,000 between them. People are moving away from areas of old housing and declining industry to rural areas or New Towns, often in the same region as their home city. Some people, preferring to live in greener, open spaces, move further away to more rural regions or resort areas.

London Docklands is an example of urban regeneration.

Inner city decline
The areas which have suffered most from population loss are the inner cities of the largest urban areas. People who can afford to, have moved out into the suburbs or further afield. Those left behind are often the poorest, including the unemployed, students and single parent families.

Since the 1970s, governments have tried to stop this movement and to regenerate the city centres and the inner city, but so far this has not had much effect. In the 1990s, however, there have been signs that inner city decline has been slowing down.

Internal migration in the UK (1981 - 1991)

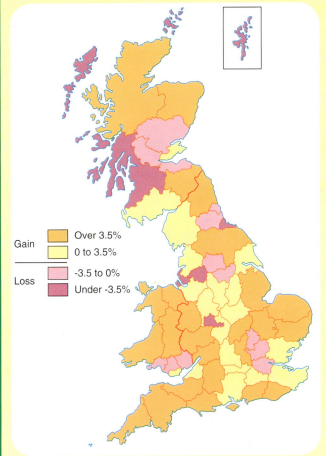

Gain
- Over 3.5%
- 0 to 3.5%

Loss
- -3.5 to 0%
- Under -3.5%

SHOULD WE RUIN THE COUNTRYSIDE OR REVIVE THE CITY?

The population movement away from British cities might be coming to an end, but no one can be sure. Two newspaper articles from October and November 1996 illustrate this point.

'The British still seek a rural idyll, and ruin it'
Every day, 300 people abandon Britain's cities. Car exhausts, soiled pavements, asthma, crime, traffic jams and poor housing make urban life unbearable. Our cities are neglected and decaying so we hate them. In one survey, 81% of people said they would prefer to live in a village or small town, but 70% of us live in large cities.

As country lovers move out of the city, the countryside is being covered with motorways, business parks, hypermarkets and housing estates. The countryside that we say we love is being obliterated whilst our cities become derelict.

'Civilising our cities'
Something is stirring in Britain's cities. Slowly but surely they are becoming desirable places to live. The stampede away from the city has become a trickle. In some big cities, London, Leeds and Manchester, the trend is being reversed and people are moving back to the urban hub. The re-colonisation is led by the affluent middle class. They are mostly under 40, single or childless couples working in professions. They like the bright lights, restaurants, theatres and, above all, lack of commuting.

STAGE 2: Liverpool's population

In Stage 1, you saw that Liverpool, like most UK cities, has lost population. In the resources that follow, there is more detailed information on this movement of people from Liverpool, together with some of the reasons.

Describe the population change in Liverpool since 1700 and make brief notes that describe and explain this change. Compare the post 1951 position in Liverpool with other UK cities (using the figures from Stage 1).

Counter-urbanisation

During the past few decades a process of **counter-urbanisation** has begun in Europe and the USA: a movement away from the larger cities into smaller towns. This is due to:

- governments trying to reduce city congestion
- offices and factories being relocated as high rents, local taxes and traffic congestion make city locations unattractive
- industrial growth in new industries outside old urban areas cause people to move to New Towns or growth areas
- local government schemes to clear slums in inner city areas
- the relatively high proportion of older people in developed countries retiring to pleasanter environments.

Five stages in urban life

1. Steady migration from countryside to city. Main growth in central area.
2. Rural-urban migration increases, fastest growth in suburbs, city begins to expand outwards.
3. Inner areas begin to lose population while the growth of suburbs continues more slowly.
4. Loss of population by the whole city as counter-urbanisation begins. People move beyond the **Green Belt** (ie, countryside that is protected from development) to New Towns and rural areas.
5. Re-urbanisation as the population of the central area stabilises. Governments promote job creation in the city and people see the advantages in living close to entertainment and other facilities in the city centre.

Liverpool population

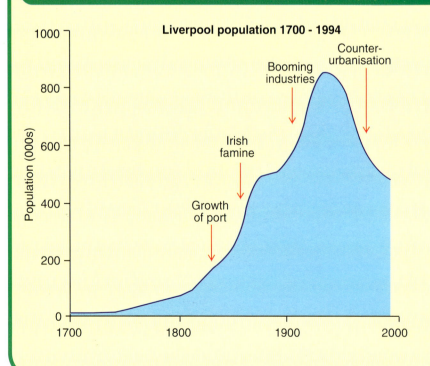

Liverpool population 1700 - 1994

(rounded to the nearest 1,000)	
1700	5,000
1720	10,000
1740	20,000
1750	40,000
1801 (first census)	78,000
1811	94,000
1831	165,000
1851	251,000
1871	493,000
1891	518,000
1911	746,000
1931	856,000
1951	789,000
1971	610,000
1991	481,000
1994	477,000

The growth of Liverpool

In the early eighteenth century, Liverpool was a small fishing port and harbour. The port began to grow rapidly in the second half of that century when huge profits from the slave trade financed new docks and industries. Woollen and cotton goods were exported all over the world while sugar and raw materials were imported.

The nineteenth century saw massive growth in Liverpool. The population rose from around 75,000 in 1800 to 750,000 in 1911 - a 1000% increase. In some decades the rise was as fast as in any less developed country today. This was during the great Industrial Revolution when Liverpool's docks expanded to cope with the huge volume of imports to supply the industries of Lancashire. Textiles (especially cotton) and engineering industries boomed and most of the exports went from Liverpool. Dock workers and factory workers flooded into Liverpool from the Lancashire countryside and further afield, notably from Ireland.

Liverpool continued to grow until the Second World War (1939 -1945). But since then, its population has almost halved. The docks and other industries have declined, partly because modern ships were too large for the Mersey Docks, but also because of a decline in the traditional industries of Lancashire. Since 1951, massive slum clearance and the replacement of war damaged property has reduced the population of the city as people have been rehoused in the suburbs and New Towns beyond the city boundaries.

Liverpool's growth

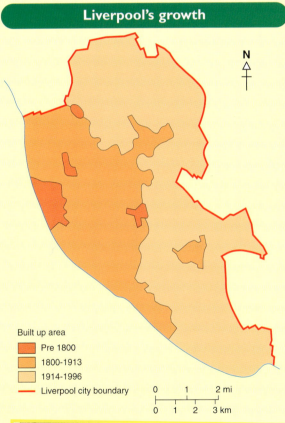

N

Built up area

- Pre 1800
- 1800-1913
- 1914-1996
- Liverpool city boundary

0 1 2 mi
0 1 2 3 km

Liverpool Pier Head - a symbol of the city's former prosperity.

Docks like this were once busy.

Georgian housing built for rich merchants.

STAGE 3: Moving out of Liverpool

So far we have seen that Liverpool grew rapidly in the nineteenth and early twentieth centuries but now its population is in decline. However, this decrease is uneven. Some parts of Liverpool have even seen a small increase in population. Other areas have seen massive **out-migration**.

In the following resources you will see the patterns of population change in different wards and some of the reasons for this pattern. Make a list of reasons why people may have left wards in the inner city.

Compare the 1928 street pattern with that in 1994. Suggest why the roads have been widened.

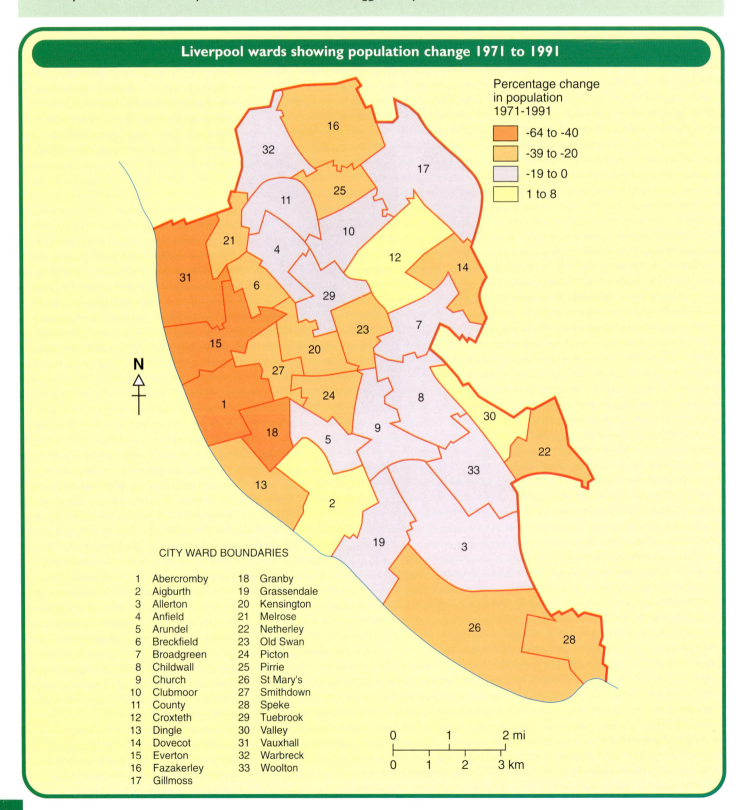

Liverpool wards showing population change 1971 to 1991

Percentage change in population 1971-1991

- ■ -64 to -40
- ■ -39 to -20
- ■ -19 to 0
- ■ 1 to 8

N

CITY WARD BOUNDARIES

1	Abercromby	18	Granby
2	Aigburth	19	Grassendale
3	Allerton	20	Kensington
4	Anfield	21	Melrose
5	Arundel	22	Netherley
6	Breckfield	23	Old Swan
7	Broadgreen	24	Picton
8	Childwall	25	Pirrie
9	Church	26	St Mary's
10	Clubmoor	27	Smithdown
11	County	28	Speke
12	Croxteth	29	Tuebrook
13	Dingle	30	Valley
14	Dovecot	31	Vauxhall
15	Everton	32	Warbreck
16	Fazakerley	33	Woolton
17	Gillmoss		

0   1   2 mi

0   1   2   3 km

Liverpool wards showing unemployment rates 1991

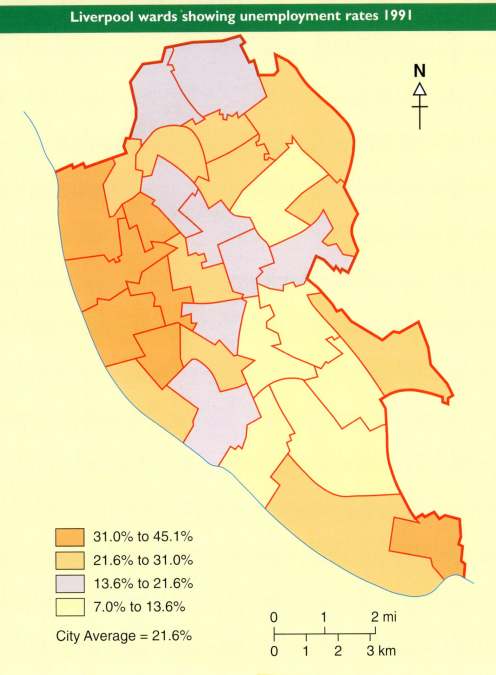

N

▮	31.0% to 45.1%
▮	21.6% to 31.0%
▮	13.6% to 21.6%
▮	7.0% to 13.6%

City Average = 21.6%

```
0        1        2 mi
|---|---|---|---|
0    1    2    3 km
```

Population change 1971 - 1991

- Liverpool lost 22% of population, this was the heaviest of any UK urban area.
- Because of an aging population there were 6,000 more deaths than births in the 1970s.
- People left to seek jobs elsewhere and, in many wards, unfit housing was demolished.
- Large scale clearance of unsatisfactory council flats and maisonettes occurred in wards to the north of the city centre and in some of the edge-of-city council estates.
- The biggest loss of jobs was in the docks and dock related industries along the Mersey, north of the city centre.

Poorly built maisonettes have been cleared.

High rise flats have been built in the city outskirts.

Dock side factories have been demolished.

Liverpool in 1928 and 1994

The Ordnance Survey map extracts show the Scotland Road area just north of Liverpool city centre. Road widening and slum clearance schemes have greatly reduced the number of people living in the area. In the north west (top left) part of the maps you can see that modern housing, with gardens, has replaced the old terrace streets.

1928 Scale 1: 10,560 (or ten inches to the mile)
Reproduced from the 1928 Ordnance Survey map

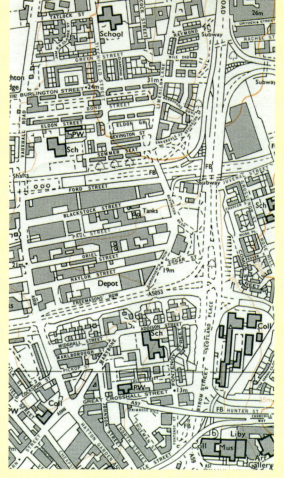

1994 Scale 1:10,000

STAGE 4: Where have all the people gone?

You have seen that the population of Liverpool has been falling for 50 years and that some areas of the city have lost population very quickly. Where have all these people gone?

The resources that follow include information on the growth of New Towns and the movement of people away from large urban areas. Make brief notes about the growth of New Towns and draw a sketch diagram showing how Skelmersdale, a New Town near Liverpool, is different from older urban areas. Finally, consider whether New Towns have been a success and make a list of advantages and disadvantages of living in one.

Moving out

The 1991 Census showed that the population of most major UK cities was falling. Between 1971 and 1991, more people left urban areas than moved in. Some reasons for this are:

- continued slum clearance in inner cities
- the growth of New Towns
- traditional industries have closed in the city centres
- new jobs are often at the city edge
- people are more mobile because of higher car ownership, so can live further from their work
- higher average incomes have allowed some people to move to bigger homes in the suburbs.

This outward migration is sometimes called **centrifugal movement**. Today, shopping malls, sports complexes and industrial estates as well as new private and council housing developments, are located on the outskirts of cities.

High rise flats in Kirkby.

In Merseyside, areas like Huyton and Kirkby grew rapidly during the 1950s and 60s. Since then there has been movement out of Merseyside to semi-rural areas of Cheshire, Lancashire, North Wales and to the New Towns of Skelmersdale, Warrington and Runcorn.

New Towns in the UK

The 1946 New Towns Act was intended to provide homes for the 'overspill' population from the inner areas of **conurbations** (ie, major cities). It was thought at the time that the population would rise rapidly and New Towns would avoid 'urban sprawl' eating into the Green Belt. Another motive for the New Towns was to attract jobs and businesses to depressed industrial areas in the North of England, Central Scotland, Northern Ireland and South Wales. Early New Towns were mainly near London. The 1960s saw a second New Town generation, more evenly spread across the country. From 1970 to 1978, 'Expanded Towns' were created using existing centres and infrastructure (roads, railways, power lines, public services). These were cheaper to develop and were based around existing 'communities'. Three New Cities with populations up to 250,000 were planned but Milton Keynes was the only totally new community.

The 1978 Urban Areas Act diverted government money away from New Towns to inner cities because:

- money was needed to regenerate the inner cities
- inner cities were beginning to have severe social problems
- New Towns near conurbations did not generate enough jobs; many residents had to commute back to the cities they had left or were unemployed
- some New Towns had attracted industries away from cities
- London's New Towns form a new commuter belt.

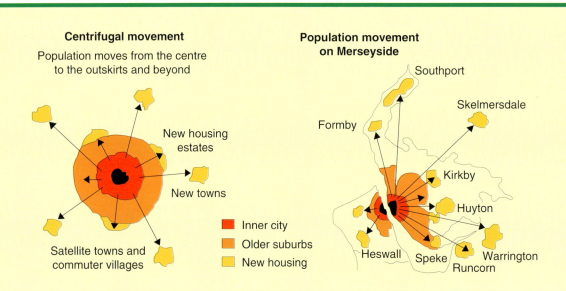

Centrifugal movement

Population moves from the centre to the outskirts and beyond

New housing estates

New towns

Satellite towns and commuter villages

- Inner city
- Older suburbs
- New housing

Population movement on Merseyside

Southport

Skelmersdale

Formby

Kirkby

Huyton

Heswall

Speke

Warrington

Runcorn

Skelmersdale

In 1961 Skelmersdale, a small town of 10,000 people in Lancashire, was designated a New Town to accommodate overspill population as redevelopment took place in Liverpool. It was estimated that 50,000 people would move there. The planners decided that:

- high private car usage must be catered for, while still encouraging public transport
- pedestrians and traffic should be separated in the town centre and residential areas
- industry should be concentrated in separate, but easily accessible areas
- open spaces should be provided but the housing should be designed to foster a close community atmosphere
- there must be no urban sprawl, and a clear boundary between town and country
- the town must try and attract a balanced population structure.

Housing in Skelmersdale is high density but there are no high rise flats. Residential areas are close to the town centre where most facilities have been concentrated.

Skelmersdale town centre. Traffic is kept away from pedestrians.

New Towns in the UK

1 Ballymena
2 Antrim
3 Craigavon
4 Glenrothes
5 Cumbernauld
6 Livingston
7 East Kilbride
8 Irvine
9 Cramlington
10 Killingworth
11 Washington
12 Peterlee
13 Aycliffe
14 Skelmersdale
15 Warrington
16 Runcorn
17 Newtown
18 Redditch
19 Cwmbran
20 Peterborough
21 Corby
22 Northampton
23 Stevenage
24 Welwyn Garden City
25 Hatfield
26 Harlow
27 Basildon
28 Hemel Hempstead
29 Bracknell
30 Crawley
31 Central Lancashire ⎫
32 Telford   ⎬ Planned new cities
33 Milton Keynes   ⎭

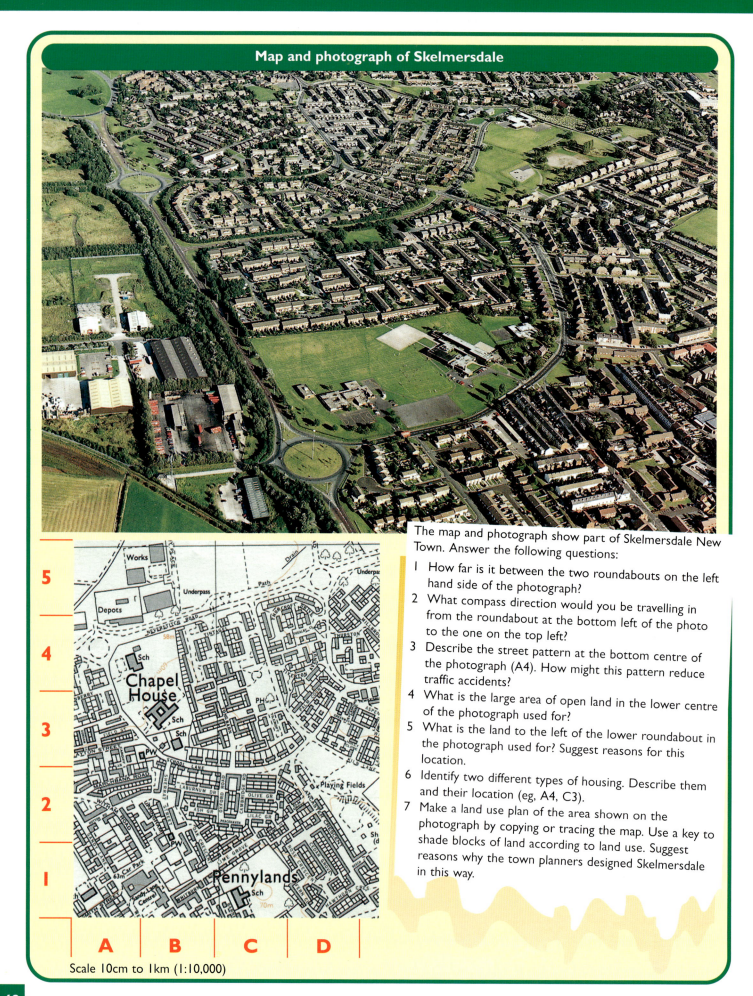

Map and photograph of Skelmersdale

Scale 10cm to 1km (1:10,000)

The map and photograph show part of Skelmersdale New Town. Answer the following questions:

1 How far is it between the two roundabouts on the left hand side of the photograph?

2 What compass direction would you be travelling in from the roundabout at the bottom left of the photo to the one on the top left?

3 Describe the street pattern at the bottom centre of the photograph (A4). How might this pattern reduce traffic accidents?

4 What is the large area of open land in the lower centre of the photograph used for?

5 What is the land to the left of the lower roundabout in the photograph used for? Suggest reasons for this location.

6 Identify two different types of housing. Describe them and their location (eg, A4, C3).

7 Make a land use plan of the area shown on the photograph by copying or tracing the map. Use a key to shade blocks of land according to land use. Suggest reasons why the town planners designed Skelmersdale in this way.

Living in Skelmersdale

Joyce Evans
In the 1950s and 60s all our family lived on the same street in Kensington (an inner suburb of Liverpool). There were my grandparents, my two aunts, my five cousins and our family all close together. Everyone used to leave their front door open during the day so we could drop in and visit whenever we liked. You can't do that now in Skelmersdale. Back in Liverpool the houses were all the same. It was a terraced street with the front door opening onto the pavement and a back yard leading onto an alley.

In the late 1960s, the council said the houses were unfit to live in and they would all be knocked down. They built some new houses nearby but they were bigger and they have gardens so not as many people live in the area as in the old days. Out of all our family, only one cousin still lives there, he was rehoused by the council in one of the new houses. Everyone else has moved out, either to the suburbs or here in Skelmersdale. It's better here, with all modern amenities but you don't feel like you know the people around you.

Bob Stanley
I was born at Everton Brow in Liverpool. The whole area used to be covered in terraced streets built in the 1880s. Nobody had a garden, there were just small back yards. In the 1950s the whole area was declared a slum by the council. Most of the people I knew moved out to Kirkby. Those people who stayed behind were rehoused in flats. Some were low rise but most were tower blocks. They were no good for families, there was nowhere safe for the children to play. The flats were poorly built and often damp. Now most of them have been knocked down and there is a big park where once so many people lived. Skelmersdale is great for bringing up kids with lots of parks and open space.

New towns sometimes appear desolate and unfriendly.

New houses in Skelmersdale with plenty of open space.

STAGE 5: Review

By now, you should have a collection of maps, graphs and notes about the changing population of Liverpool. Your final task is to write an essay, using your notes, illustrated by your maps and graphs.
The title of the essay is: 'Why are people leaving Liverpool?'

In an essay like this you need to make sure that you are thorough and accurate, but do not include information that is irrelevant. Your essay can be written in a number of ways but it should include some information about the following:

- the pattern of population change, when did people leave, which areas are they leaving?
- what are the reasons for leaving, and what are the attractions of the new places to which they are moving?
- is it likely that present trends will continue and, if they do, what will happen to Liverpool and similar cities?

Glossary

All the terms listed below are explained in this Enquiry. In each case, write a definition of the term and, if possible, give at least one example.

Counter-urbanisation	**Centrifugal movement (of population)**
Census	**New Town**
Urban regeneration	**Conurbation**
Green Belt	**Overspill**
Out-migration	**Satellite town**

4 Leisure and tourism

to the student

The rapid growth of mass market travel over the past 40 years has made tourism the biggest business in the world. In 1995, international tourism earned $372 billion and over 500 million people travelled across their national frontiers.

The growth in tourism has partly come about through increases in people's incomes and increased leisure time, especially in the more economically developed countries (MEDCs). There is greater mobility as charter flights and package tours have opened up a wide range of convenient and cheaper foreign holidays. Tourism brings economic wealth to the holiday destinations. However, it can also be a threat to local environments and cultures if uncontrolled development is allowed.

In the Enquiries that follow, we shall investigate global patterns of tourism and, in particular, the impact of tourism on:
- Corfu, a long established holiday spot in a MEDC
- Sri Lanka, one of the fast developing less economically developed countries (LEDCs)
- the Northumberland National Park, a UK leisure area.

questions to consider

1 What are the trends in international tourism?
2 Why is 'package tourism' successful?
3 What is the impact of tourist development on the economy, environment and culture of destinations in MEDCs and LEDCs?

key ideas

Mass tourism - as average incomes have increased in MEDCs, millions of people now take advantage of cheap charter flights and package holidays to take holidays abroad.

Package holidays - travel and accommodation are arranged by a tour company. The alternative is for individuals to make their own arrangements but, because they do not have the bargaining power of the tour companies, they pay more.

Honeypots - sites of scenic or historic interest that attract large numbers of visitors. These sites can be so overwhelmed by visitors that they lose their original appeal. Sometimes the term *hotspot* is used to describe any popular tourist destination.

Sustainable development, in tourism, means that resources are managed in such a way that the original attraction is not destroyed. In other words, a balance is kept between economic development and conservation of the natural landscape.

activities

Using information from this and the facing page:

a) Suggest reasons why mass tourism has increased.
b) Working with a partner, describe ways in which tourism can impact on an area. In each case, write a sentence that explains the point.

RISING INCOMES → INCREASED LEISURE TIME → CHEAPER TRANSPORT

LEISURE AND TOURISM
The World's Biggest Business

BUSINESS

TRADITIONAL PACKAGE

LONG HAUL LOCATIONS

NATIONAL HERITAGE

Going Places

VANISHING TRADITIONS AND CULTURES ← LAND USE CONFLICTS ← JOBS

Enquiry

Eurotourism

In this Enquiry you will investigate Mediterranean 'package' tourism. Modern mass tourism started in the Mediterranean in the late 1950s and early 1960s. The Spanish coast north of Barcelona (known as the Costa Brava) and the Greek Islands (such as Corfu) were among the first successful resort areas. You will be given information on a particular package holiday destination - Corfu - and its attractions. In addition, the impact of tourism will be described and you will be asked to consider the advantages and disadvantages of tourist development.

The outcome of this Enquiry will be an 'alternative' tourist guide to Corfu that you will be asked to produce. It could be in the form of a booklet or information sheet and should include maps, tables and written information from the Stages of your Enquiry.

STAGE 1: Where are the Mediterranean package tour hotspots?

You are asked to prepare a map of the Mediterranean showing the most popular destinations, or hotspots, for UK holidaymakers. List the destinations and note the countries in which they are located. Suggest reasons why Mediterranean resorts are popular.

The list of Mediterranean countries shows those which received the most tourist visitors in 1996. France is included on the list because it has a Mediterranean coastline. However, it is not a major package tour destination and most of its visitors did not visit the southern coast.

Make a note of which country had the biggest growth in tourism (between 1994-96) and suggest possible reasons for this.

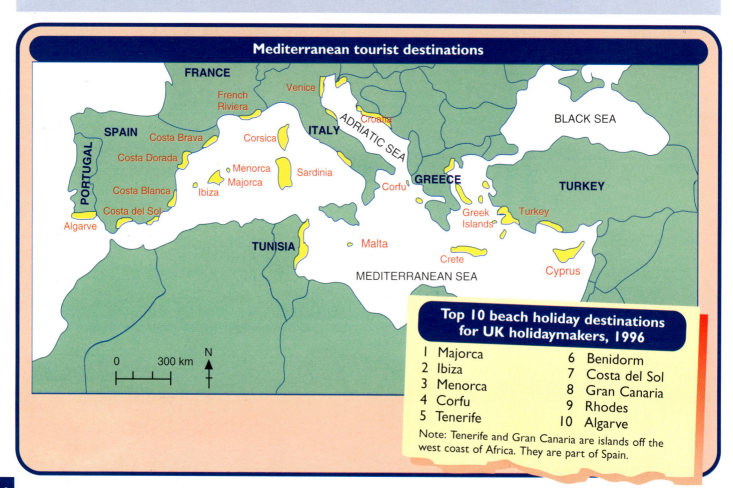

Mediterranean tourist destinations

Top 10 beach holiday destinations for UK holidaymakers, 1996

1	Majorca	6	Benidorm
2	Ibiza	7	Costa del Sol
3	Menorca	8	Gran Canaria
4	Corfu	9	Rhodes
5	Tenerife	10	Algarve

Note: Tenerife and Gran Canaria are islands off the west coast of Africa. They are part of Spain.

Majorca

With sleepy villages and busy resorts, sheltered coves and broad beaches, lively nights and historic sights, there's a perfect place for every taste in Majorca.

3 star hotel, 7 nights in July, £479 per person.

Costa Blanca

The largest resort is Benidorm. It's lively with a British feel and all the familiarity of home, together with the benefit of blue skies and clear blue seas.

3 star hotel, 7 nights in July, £369 per person

Tunisia

Somewhere a little different, a taste of Africa on Europe's doorstep. Long sandy beaches and historic old towns with their narrow streets and bazaars.

3 star hotel, 7 nights in July, £449 per person.

Algarve

Excellent beaches, golf courses and an average 300 cloudless days a year. The Algarve is a regular favourite amongst family holidaymakers.

3 star hotel, 7 nights in July, £429 per person.

Corfu

Hot summer months and a friendly population combine to make Corfu a family favourite. Good hotels and lively towns ensure there is something for every taste.

3 star hotel, 7 nights in July, £499 per person.

Menorca

The ideal place for a relaxing family holiday with a hundred sandy bays. The resorts are large enough for entertainment yet small enough to stroll around.

3 star hotel, 7 nights in July, £449 per person.

Turkey

A land at the crossroads of different cultures. Turkey has some of the most beautiful, unspoilt beaches in the Mediterranean and offers the best value around.

3 star hotel, 7 nights in July, £379 per person.

International tourist arrivals for selected Mediterranean countries (and Portugal), 1996		
Country	Number of visitors (millions)	% change from 1994
France	60.6	-1.2
Spain	45.1	4.4
Italy	29.2	6.2
Greece	11.1	3.6
Portugal	9.5	4.2
Turkey	6.5	7.9
Tunisia	4.1	7.0

Note: France is included on this list because it has a Mediterranean coastline. However, it is not a major package tour destination and the Mediterranean is only one of its many tourist areas.

STAGE 2: What is it like on holiday in Corfu?

You have seen in Stage 1 that Corfu is a popular tourist destination. In the resources that follow you will find reasons why, despite competition from recently developed destinations such as Turkey, Corfu remains a holiday hotspot. Use the resources to write **two** postcards as though you are a holiday-maker in Corfu. On the first, write as though you are with a group of teenagers writing to friends back home. On the second, write as though you are a retired couple sending a card back to their family.

The postcards should give reasons why the tourists chose Corfu and should include descriptions of the activities and attractions they have enjoyed.

Corfu and mainland Greece attract tourists to sites of historic interest as well as to beaches.

Flights to Corfu (Kerkyra Airport) by one UK tour operator

GATWICK (3hrs 15min)
Mon 21.25   Fri 20.55

STANSTED (3hrs 15min)
Mon 23.55

BRISTOL (3hrs 30min)
Mon 10.45

CARDIFF (3hrs 30min)
Mon 21.00

BIRMINGHAM (3hrs 30min)
Mon 19.50

EAST MIDLANDS (3hrs 30min)
Mon 06.00

MANCHESTER (3hrs 30min)
Mon 21.45   Fri 22.30

NEWCASTLE (3hrs 45min)
Mon 18.30

GLASGOW (4 hrs)
Mon 21.05

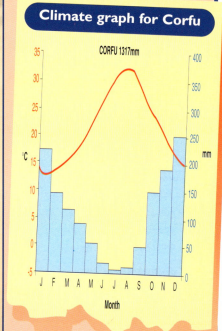

Climate graph for Corfu

CORFU 1317mm

Holiday resorts in Corfu

SIDARI
Small resort on beautiful coastline of bays and headlands.

RODA / ACHARAVI
Safe beaches. Quiet.

KASSIOPI
Attractive, fairly quiet resort with fishing boats. Mainly self catering.

PALEOKASTRITSA
Spectacular coast and clear blue water. Very quiet with scattered accommodation in olive groves.

IPSOS / DASSIA
The 'Golden Mile' of nightlife. Narrow beach with watersports.

CORFU
CAPITAL. Historic town with shops.

BENITSES
Busiest resort. Lively nightlife. Narrow shingle beach. Many watersports.

AGHIOS GORDIOS
Quiet resort with a good beach.

MORAITIKA / MESSONGHI
Quiet resorts, narrow beaches.

AGHIOS GEORGIOS
Two of the best sandy beaches in the island. Quiet.

KAVOS
Fast growing resort. 3km sandy beach. High energy nightlife.

N

0   20 km

AG. GEORGIOS PAGI CORFU

IPSOS CORFU

SIDARI CORFU

STAGE 3: What is the impact of tourism in Corfu?

Satisfied holiday makers make return visits and bring spending money to the island. This helps to create jobs and funds further tourist development. The income from tourism is used to improve services such as transport, health, energy and water supply - so improving the standard of living for everyone on the island.

However, tourism has changed the island and some developments have not been welcomed by everyone. New hotels, crowded beaches, noisy discos and the loss of local culture are often blamed on mass tourism. Gerald Durrell, a famous writer who grew up in Corfu in the 1930s, wrote about the island before the time of mass tourism. An extract from his writing is in the resources that follow. Contrast the mood and image that he conveys with the material taken from recent tourist brochures.

Compare the recent photos of Corfu (from Stage 2) with the one that illustrates the Gerald Durrell extract. List the features of Gerald Durrell's Corfu which still exist.

To record the impact of tourism on Corfu, you are asked to complete a summary chart or table. Use the information provided below together with the resources from Stage 2. Set it out in the following way:

	Before tourism	After tourism
Employment		
Environment		
Settlements		

Under your chart or table write a brief conclusion. To what extent, in your view, has tourism benefited the local people?

Extract from Gerald Durrell 'My family and other animals'

The sea was smooth, warm, and as black as black velvet, not a ripple disturbing the surface. The distant coastline of Albania was dimly outlined by a faint reddish glow in the sky. Gradually, minute by minute, this glow deepened and grew brighter, spreading across the sky. Then suddenly the moon, enormous, wine-red, edged herself over the fretted battlement of mountains, and threw a straight, blood red path across the dark sea.

We returned through the sun-striped olive groves where the chaffinches were pinking like a hundred tiny coins among the leaves. Yani, the shepherd, was driving his herd of goats out to graze. His brown face, with its great sweep of nicotine stained moustache, wrinkled into a smile; a gnarled hand appeared from the heavy folds of his sheepskin cloak and was raised in salute.

With March came the spring, and the island was flower filled, scented, and a-flutter with new leaves. Waxy yellow crocuses appeared in great clusters, bubbling out among the tree-roots and tumbling down the banks.

So, we hurried down the hillside until we reached the little bay, empty, silent, asleep under the brilliant shower of sunlight. We sat in the warm, shallow waters, drowsily, and I delved in the sand around me.

Across the mouth of the bay a sun-bleached boat would pass, rowed by a brown fisherman in tattered trousers, standing in the stern and twisting an oar like a fish's tail.

First published in 1956 by Hart Davis. Reproduced here by permission of HarperCollins.

Traditional farming.

Changes to the Corfu economy

Old Corfu.

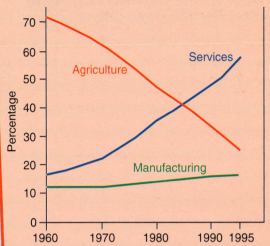

The changing employment structure in Corfu since 1960

On account of its climate, natural beauty and international airport, Corfu has developed a tourist industry that is the most dynamic part of its economy. The long term decline in population, caused by migration to mainland Greece, has been reversed. However, the depopulation of inland villages and the gradual abandonment of farming has continued. There is an internal migration of people to the main towns to work in the tourist industry. This is an example of rural - urban migration.

Agriculture is marked by low productivity, concentrating mainly on growing citrus fruit and olives, and in herding sheep and goats. A small area of irrigated land is used to produce some fresh food for the island's hotels and restaurants.

The only industrial units that operate on the island are struggling to survive. Craft industries, however are doing quite well, and here the influence of tourism is clear. One important group of small businesses is engaged in producing tourist goods, whilst another processes products from the primary sector.

STAGE 4: Review

In order to help you summarise the important issues relating to mass tourism in Corfu, you are asked to produce an 'alternative' guide to the island. This can be in the style of a 'Rough Guide', in other words it should be an honest view of what a holidaymaker might expect. Produce the guide as a booklet or information sheet(s). Include any maps, diagrams, statistics or written extracts that you think will be helpful. Use the information that has been provided plus any other that you can gather.

Remember that the point of this exercise is not simply to produce a 'glossy' brochure that only shows the good points. Your guide should include, for example, some comment on whether - in your view - tourism has 'spoilt' the island.

LIVERPOOL HOPE UNIVERSITY COLLEGE

enquiry

The impact of tourism in an LEDC - Sri Lanka

In the previous Enquiry, you looked at package tourism in the Mediterranean. In the 1990s, the popularity of the favourite Mediterranean holiday playgrounds is being challenged by 'long haul' destinations. Some of these are in countries like the USA and Australia but others are in Less Economically Developed Countries (LEDCs) such as Malaysia, Kenya and Sri Lanka. Many tourists are looking for a change from the increasingly familiar beaches of the Mediterranean. They are keen to travel greater distances to more exotic and varied destinations.

Sri Lanka is an example of an LEDC which has experienced the rapid rise of a tourist industry. In this Enquiry you will investigate the reasons for Sri Lanka's tourist development and the impact that it has had.

The outcome of the Enquiry will be a report written by you in the role of a government official in Sri Lanka. Your report will outline recent trends in tourism and make the case for further, government funded, development.

STAGE 1: What are the current trends in world tourism?

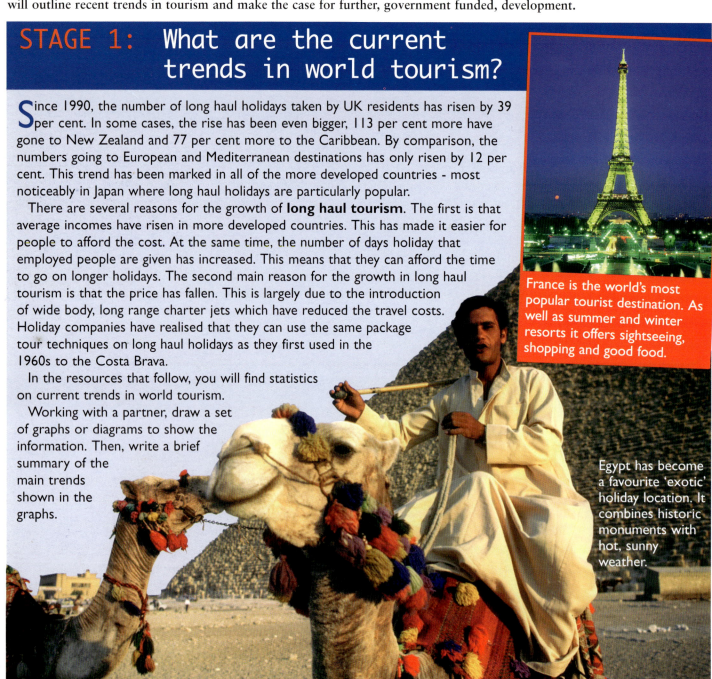

Since 1990, the number of long haul holidays taken by UK residents has risen by 39 per cent. In some cases, the rise has been even bigger, 113 per cent more have gone to New Zealand and 77 per cent more to the Caribbean. By comparison, the numbers going to European and Mediterranean destinations has only risen by 12 per cent. This trend has been marked in all of the more developed countries - most noticeably in Japan where long haul holidays are particularly popular.

There are several reasons for the growth of **long haul tourism**. The first is that average incomes have risen in more developed countries. This has made it easier for people to afford the cost. At the same time, the number of days holiday that employed people are given has increased. This means that they can afford the time to go on longer holidays. The second main reason for the growth in long haul tourism is that the price has fallen. This is largely due to the introduction of wide body, long range charter jets which have reduced the travel costs. Holiday companies have realised that they can use the same package tour techniques on long haul holidays as they first used in the 1960s to the Costa Brava.

In the resources that follow, you will find statistics on current trends in world tourism.

Working with a partner, draw a set of graphs or diagrams to show the information. Then, write a brief summary of the main trends shown in the graphs.

France is the world's most popular tourist destination. As well as summer and winter resorts it offers sightseeing, shopping and good food.

Egypt has become a favourite 'exotic' holiday location. It combines historic monuments with hot, sunny weather.

The top ten long haul destinations for UK tourists, 1995

1 Florida
2 Dominican Republic
3 Egypt *
4 Kenya
5 Barbados
6 Antigua
7 Jamaica
8 St Lucia
9 Cancun - Mexico
10 Thailand

* Although, strictly speaking, Egypt is a Mediterranean country, the tour operators classify it as long haul.

World's Top 20 tourism destinations

Country	International tourist arrivals (millions)	
	1990	1995
1 France	52.5	60.6
2 Spain	37.4	45.1
3 USA	39.5	44.7
4 Italy	26.7	29.1
5 China	10.5	23.3
6 United Kingdom	18.0	22.7
7 Hungary	20.5	22.0
8 Mexico	17.1	19.8
9 Poland	3.4	19.2
10 Austria	19.0	17.7
11 Canada	15.2	16.8
12 Czech Republic	7.2	16.6
13 Germany	17.0	14.5
14 Switzerland	13.2	11.8
15 Greece	8.8	11.1
16 Hong Kong	6.5	9.5
17 Portugal	8.0	9.5
18 Malaysia	7.4	7.9
19 Singapore	4.8	7.9
20 Thailand	5.3	6.5
Total (top 20)	**338.5**	**415.7**
World total	**459.2**	**567.0**

The USA has slipped from second to third as a world tourist destination. Nevertheless, cities like San Francisco remain very popular with international tourists.

World Tourism Organisation forecast

International Tourist Arrivals
1995:   567,400,000
2000:   702,000,000
2010:   1,018,000,000
Tourism receipts represented more than 8% of the world's merchandise exports and 33.3% of the world trade in services in 1995.

Sri Lanka tourism is supported by the government but, between 1983 and 1990, the industry was disrupted by ethnic unrest and the violent activities of the separatist Tamil Tigers. The government has generally been successful in confining the conflict to the north east of the country. This has allowed it to encourage tourist development in the south and west. Like most LEDCs, Sri Lanka is keen to raise its earnings by attracting foreign tourists. The tourists spend money on food, accommodation and entertainment. This provides jobs and increases the incomes of those directly involved. It also indirectly helps many other people who benefit from the increased incomes of the hotel and tourist industry workers.

Unfortunately, for Sri Lanka and other LEDCs, political instability and tensions can quickly persuade long haul tourists to choose an alternative destination. At the same time, within the LEDCs, there are many competing claims for development money. For example, money can be used to build a fertiliser plant, a new road or to spend on arms and military equipment for the army. Nevertheless, in Sri Lanka, the tourism industry has managed to grow and to overcome these problems.

Using the resources that follow, make a list of the people and organisations in Sri Lanka who benefit from tourism.

Graph showing the importance of the tourist industry to Sri Lanka

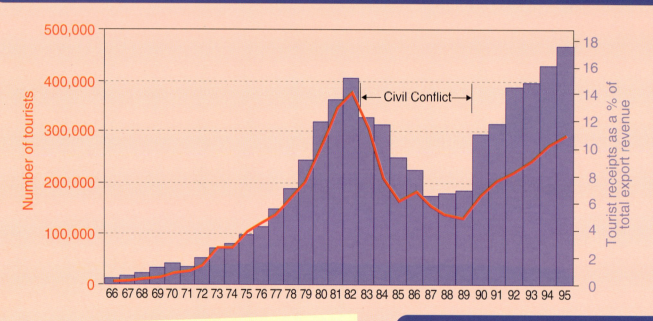

'The pearl of the Indian Ocean'

Located where the seaways east and west meet, this paradise isle has a recorded history of over 2,500 years, and boasts golden beaches, verdant highlands, wildlife sanctuaries, ancient cities and a hospitable people. An hour by helicopter or four by road separates the tropical beaches from the cool climes of the hill country.

Hotels tend to be simple, entertainment is unsophisticated, service a little slow but prices are amazingly low and everyone is so friendly.

3 star hotel, 7 nights in July, £820 per person.

- *Sri Lanka tourist brochure*

Climate graph for Sri Lanka

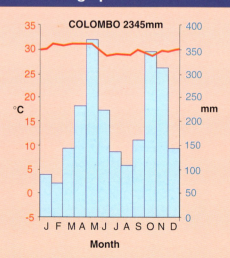

Map showing main resort area

Jaffna

Trincomalee

INDIAN OCEAN

✈ • Kandy

• Colombo

Main beach resort area →

• Galle

N
↑
0   100 km

Sri Lanka

Population - 18 million

Area - 25,332 sq. miles (about the size of Ireland)

Largest city - Colombo (population - 800,000)

Life expectancy - male: 69 years, female: 74 years

Literacy rate - 90%

GDP per capita US$640

GDP grew by 5.4% per year between 1990 and 1994 (UK grew by 0.8% per year)

Labour force - Agriculture: 48%, Industry: 21%, Services: 31%

Numbers directly employed in the tourist industry (1996): 36,000; numbers indirectly employed in tourism: 51,000; total 87,000. In 1991 the total was 64,000.

Main exports - tea, rubber, coconuts, textiles, cinnamon, ceramics

Main foreign earnings (other than exports) - tourism and remittances (money sent home by migrant workers abroad)

Terrain - Coastal plains, hills and mountains inland

Climate - Tropical (hot and wet seasons, April to June and September to November; Hot and drier seasons for the rest of the year)

Hotels provide employment for the people who work in them. They also benefit the local economy through the taxes they pay and the produce they buy. Tourists buy souvenirs from markets and handicraft stalls.

Safe tropical beaches are the main attraction for holidaymakers in Sri Lanka.

STAGE 3: Review

Using the information that has been provided, you are asked to summarise the important issues facing Sri Lanka's tourist development. Take the role of a government official. Write a report that outlines the current trends in tourism and then sets out the case for government assistance to the tourism industry. Use, where appropriate, visual material such as maps, graphs or diagrams to illustrate your points.

enquiry

Conflicting land use in the Northumberland National Park

This Enquiry uses a British example of a National Park to show how *conflicts in land use* can arise. You will see how different groups of people wish to use one particular area of outstanding natural beauty. It attracts tourists who wish to see the fragile environment conserved. However, at the same time, it contains resources which can be developed to provide employment and incomes for local people. Planning controls are one way in which the conflict between conservationists and developers can be lessened. The aim is to manage developments so that a balance is maintained between the needs of the local economy and conserving the natural landscape. This is called *sustainable development*. In practice, it is difficult to achieve.

Sometimes, when difficult planning decisions are required, a public consultation is held. This means that the public are invited to comment. The outcome of this Enquiry will be a response from you, in the form of a letter, to a public consultation exercise. You will be asked to write whether, from your point of view, permission should be granted for gravel extraction on land just on the National Park boundary.

STAGE 1: What are the conflicting land use claims?

In 1996, a company called Northern Aggregates sought permission to extract sand and gravel from the Ingram Valley, just outside the Northumberland National Park. Some local people who wished to protect this area of outstanding natural beauty protested.

Write a brief note that describes what National Parks are, where they are situated and why they should be protected. Use the resources to identify the different points of view of the people involved in the land use conflict. It might be helpful to discuss these with a partner or, if possible, have a class discussion of the issues. You should have your own notes that list the people and organisations involved, what they want and what conflicts might arise.

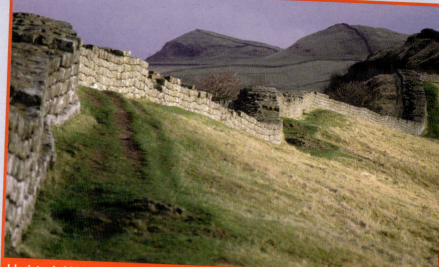

Hadrian's Wall lies at the southern end of the Northumberland National Park. The Park mainly consists of wild, open moorland. It includes a number of sites which are of important archaeological value.

National Parks

There are 12 National Parks in England and Wales. Their aim is to provide protection for the outstanding countryside that they contain and, secondly, to provide opportunities for access and outdoor recreation. They are 'national' in the sense that they are of value to the nation as a whole. Most of the land is privately owned.

Special planning regulations control development in the National Parks. This is also the case in Scotland for the four regional parks and 40 National Scenic Areas.

In England and Wales, in addition to the National Parks, there are 40 Areas of Outstanding Natural Beauty which have scenic value but lack extensive areas of open countryside suitable for recreation and, therefore, National Park status. Nevertheless, local authorities maintain strict planning controls over these areas.

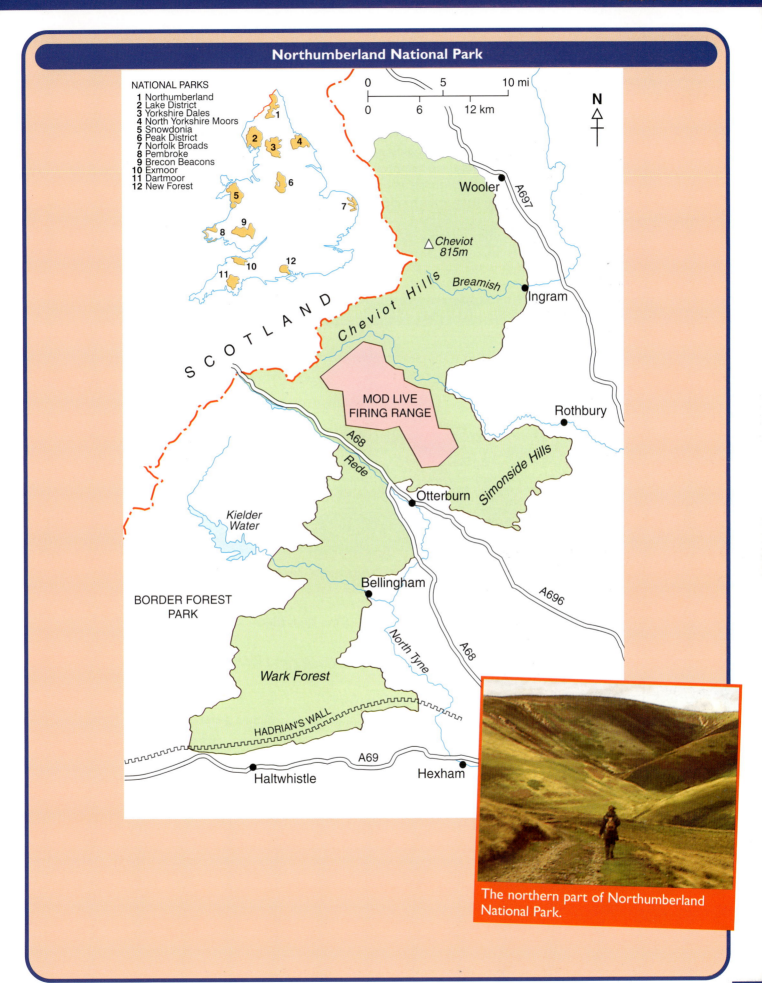

Northumberland National Park

NATIONAL PARKS
1 Northumberland
2 Lake District
3 Yorkshire Dales
4 North Yorkshire Moors
5 Snowdonia
6 Peak District
7 Norfolk Broads
8 Pembroke
9 Brecon Beacons
10 Exmoor
11 Dartmoor
12 New Forest

| 0 | 5 | 10 mi |
| 0 | 6 | 12 km |

N

SCOTLAND

Wooler

Cheviot Hills

△ Cheviot 815m

Breamish

Ingram

MOD LIVE FIRING RANGE

Rothbury

A68

Rede

Simonside Hills

Otterburn

Kielder Water

BORDER FOREST PARK

Bellingham

North Tyne

A696

A68

Wark Forest

HADRIAN'S WALL

A69

Haltwhistle

Hexham

A697

The northern part of Northumberland National Park.

STRIVE - Save The Real Ingram Valley Environment

This is a campaign group which opposed the planning application that Northern Aggregates made to extract gravel. They made the following points:

1. The proposed extraction will remove 1.5 million tonnes of sand and gravel over a 10 year period.
2. The valley is at the entrance to the National Park. It will spoil the view and disturb walkers, cyclists, bird watchers and wildlife enthusiasts. Many local people depend on tourism for their livelihood.
3. The workings will be an eyesore and the restoration will completely alter the character of the landscape.
4. The natural flow of local rivers might be seriously impaired by the excavation.
5. Quarries and the proposed lakes are dangerous to children. Every year, young people are drowned when playing near open water pits.
6. Over 50 lorries per day will be needed to carry the gravel and sand. These will pass along narrow country roads by houses and small hamlets. The noise, dust and danger are unacceptable.
7. Some of the best agricultural land in the area will be lost by the excavation.

The case from Northern Aggregates

The flat ground in the foreground is the area that will be excavated for sand and gravel if the planning application is successful.

The Ingram Valley contains a valuable natural resource - over 1.5 million tonnes of sand and gravel - which is needed for the construction industry.

If we do not use local resources, we shall need to transport sand and gravel for longer distances, so increasing its cost.

New jobs will be created locally for a period of at least 10 years.

The new jobs will have a 'knock on' effect on the local economy - there will be a general rise in incomes as the demand for goods and services increases.

The site will be completely restored after the extraction. Lakes will provide wildlife habitats and opportunities for angling or fish farming.

STAGE 2: Review

When a company wishes to start excavating sand and gravel, it must obtain planning permission from the local authority. There is then a consultation period in which members of the public and other interested bodies can express their views. Using the information provided in Stage 1, write a letter to the Planning Officer on the issues raised for public consultation, ie should permission be granted to Northern Aggregates to extract gravel from the area?

Begin by stating the main facts of the planning issue. Then, explain whether, in your view, the proposal should be accepted or rejected.

Open cast excavation for sand and gravel can bring economic benefits as well as causing environmental costs.

The County Planning Office
Northumberland County Council

To members of the public

Re: Planning application for gravel extraction in the Ingram Valley.

A planning application has been made by Northern Aggregates to extract 1.5 million tonnes of sand and gravel in the Ingram Valley.

You are invited to comment on this planning application.

Yours faithfully

County Planning Officer

Glossary

All the terms listed below are explained in these Enquiries. In each case, write a definition of the term and, if possible, give at least one example.

Mass tourism	**Long haul tourism**
Package holiday	**Land use conflict**
Honeypot	**National Park**
Sustainable (tourist) development	

5 Development studies

to the student

One definition of *development* is 'growth'. Development occurs when people's income and wealth increase. This might be on a local, national or global scale. An unequal distribution (ie, sharing) of wealth leads to contrasts in people's standard of living and quality of life.

The United Nation's Human Development Report of 1996 voices concern that 'the global gap between rich and poor is widening every day, both between countries and within countries'.

In this Enquiry you will see how a range of economic and social *indicators* can be used to measure development. These include, for example, average income, life expectancy and literacy rate. You will also investigate contrasts in standards of living between different countries and regions. Later in the Enquiry, we shall consider how *economic development* takes place and how international aid can help raise standards of living.

The example of Tanzania, a less developed country, is used to show how development can be measured and encouraged. It illustrates how aid programmes can help growth and how this affects the quality of life of the population.

questions to consider

1 How can the level of development be measured?
2 Are there differences in the levels of development within the UK?
3 What is the global pattern of development?
4 How can development be encouraged?
5 How effective are international aid programmes in raising levels of development?

key ideas

Economic development is the process of increasing the output, income and wealth of people in a country or region. *Social development* occurs when people's standard of living as a whole increases.

Development indicators are statistics which are used to measure the level of development. They can include economic indicators such as National Income and social indicators such as the literacy rate.

Gross National Product (GNP) is the total value of goods and services produced within a country in one year. It is one measure of a country's *National Income*. When GNP is divided by the number of people in the population it is called *per capita GNP* and is a measure of average income. Very broadly, countries can be divided into *MEDCs* (more economically developed countries - with high per capita GNP) and *LEDCs* (less economically developed countries - with low per capita GNP). On a world scale, most MEDCs are in the North and most LEDCs are in the South.

activities

Using information from this and the facing page:

Working with a partner, 'brainstorm' a list of differences (ie, quickly think of as many as you can) between the richer, developed North and the poorer, less developed South.

WHAT IS DEVELOPMENT?

Mechanised Food Production

Healthcare and Social Services

AMBULANCE

Modern transport and communications

RICH NORTH
GNP per capita US$ 23,420 (in 1994)

North Africa and Middle East ($1,580)

South Asia ($320)

East and SE Asia ($820)

Latin America and the Caribbean ($3,340)

Sub-Saharan Africa ($460)

POOR SOUTH
GNP per capita

Hard manual work in the fields

Difficult lives with little money or time for leisure activities

Slow and inefficient transport

enquiry

Why is development important?

Development creates wealth and, if it is shared out, reduces poverty. Most people believe that rising incomes lead to a higher standard of living. For those on the poverty line this means more to eat and better shelter and health care. For higher income groups it means more *consumer goods* such as cars and televisions and also more *services* such as holidays and meals out. It is generally accepted that higher incomes allow people to enjoy a higher quality of life.

Although the world is sometimes divided into the rich North and the poor South, the true facts are more complex. Within rich countries there is poverty - for example, in some inner cities - and there are people with high incomes within poor countries. Nevertheless, it is true to say that most of the people in the world who are living in poverty are living in the less developed countries of the South.

In the Enquiry on Population (see pages 4 - 25), the link between low incomes and rapid population growth is made clear. It seems that the best way to slow population growth is to raise the incomes and level of development of people in Africa, Asia and Latin America. (Latin America is South and Central America.)

This Enquiry looks at development, what it is, where it is occurring fastest and how it can be encouraged. The outcome of the Enquiry will be an article that you will write. It will be in the form of an information leaflet from an aid agency such as Oxfam. You will be asked to summarise issues such as the definition and measurement of development; the importance of development and how it can be achieved in less economically developed countries.

STAGE 1: What is the global pattern of development?

The most widely used **indicator** (ie, measure) of economic development is per capita GNP. Although, in broad terms, the world can be divided into the rich North and the poor South, this does not really give the full picture. The map of world GNP on the facing page shows a more detailed pattern. You will see that there are some countries in the South which are much richer than others. The United Nations and the World Bank list some countries as being 'middle income' rather than simply high or low income.

Use the map to write a summary of where the highest and lowest income countries are located. Briefly note whether the simple rich North - poor South divide of the world gives a true picture of reality. Bear in mind the captions on the photographs on this and the facing page.

Using the information in the table, describe the trends that occurred in world GNP from the mid 80s to the mid 90s.

We should always remember that in high income countries there are millions of poor people, and in low income countries there are millions of rich people. In the USA, one of the richest countries in the world, there are many poor people.

Average annual growth in per capita GNP (%) 1985 - 1994

Low and middle income economies

Sub-Saharan Africa *	-1.2
East Asia and Pacific	6.9
South Asia	2.7
North Africa and Middle East	-0.4
South and Central America	0.6

High income economies

Japan, North America, Australasia and Europe	1.9

* Sub-Saharan Africa is the whole of Africa south of the Sahara Desert.

Although India has a very low average GNP it has millions of people who enjoy a high standard of living.

World Gross National Product per capita in US$ (1995)

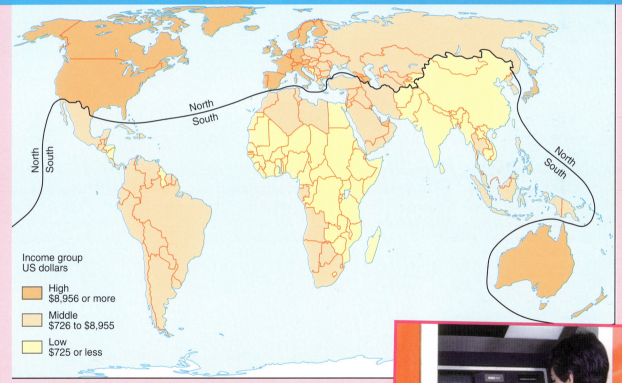

Income group
US dollars

High
$8,956 or more

Middle
$726 to $8,955

Low
$725 or less

The most commonly used means of comparing the level of development in different countries is to use per capita GNP.

Development raises people's incomes so they can afford to buy more consumer goods and services.

STAGE 2: How is development best measured?

In 1994, the per capita GNP (expressed in US dollars) of the UK was $18,340 and in Tanzania it was only $140. It is clearly the case that the UK is a much richer country than Tanzania, but just how much richer?

Many people in Tanzania grow and eat their own food so they do not have any money income at all. Also, in Tanzania, $10 buys more than in the UK, so GNP exaggerates the poverty gap. On its own, per capita GNP does not tell the full story. Social indicators such as life expectancy must also be used to measure and compare standards of living and social development.

In the resources that follow you are provided with a range of indicators for the UK, Tanzania and other selected countries. Because any one indicator might give a misleading impression, international bodies like the United Nations produce 'composite indicators'. One example is the 'Human Development Index' which combines income, education and life expectancy. Compare the GNP map from Stage 1 with the HDI map below. Make a brief written note of where they appear to differ.

You are asked to construct your own index (following the instructions provided) and write down the selected countries in rank order. Then, briefly describe what your index shows and comment on how true a picture you think it gives.

The Human Development Index (HDI)

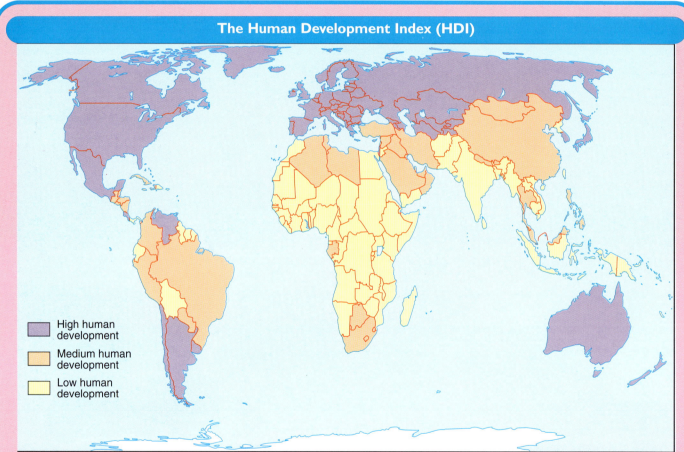

High human development

Medium human development

Low human development

The Human Development Index was devised in 1990 by the United Nations Development Programme (UNDP) as an alternative method of measuring development. It is based on three indicators including both social and economic factors:

- income per capita adjusted for different costs of living
- educational attainment measured by adult literacy and average number of years of schooling
- life expectancy at birth.

The combined score of the 3 indicators is expressed as a value between 0 and 1. The nearer to 1, the higher the level of development.

The UK HDI score is 0.916.

The Tanzania HDI score is 0.364.

Development Indicators for 20 selected countries (1994)

	GNP per capita (US$)	Average life expectancy	Adult literacy (%)	Infant mortality (per 1,000 births)
Australia	17,510	77	99	7
Brazil	3,020	66	82	57
Burkino Faso	300	49	17	127
China	490	71	79	27
Egypt	660	61	49	57
France	22,360	77	99	7
Germany	23,560	75	99	8
India	290	60	50	88
Italy	19,620	76	97	9
Japan	31,450	79	99	5
Malaysia	3,160	71	82	20
Mexico	3,750	70	87	36
Nigeria	310	52	53	96
Saudi Arabia	8,000	66	61	58
Somalia	500	47	27	122
South Africa	2,900	63	81	62
Taiwan	11,000	74	93	6
Tanzania	140	51	64	97
United Kingdom	18,340	76	99	8
USA	24,750	76	99	8

Standards of living depend on more than money income. Education and health care are also important.

How to construct an index of development

Using a copy of the 20 country table, work out and write down the rank number for each indicator. Rank number 1 being the highest level of development. For example:

	GNP per capita (US$)	(Rank)	Average life expectancy	(Rank)	Adult literacy (%)	(Rank)	Infant mortality (per 1,000 births)	(Rank)	Total of rank orders
Australia	17,510	(7)	77	(2=)	99	(1=)	7	(3=)	13

Then rewrite the list of countries in overall rank order with the lowest scoring country at number 1. This is the most developed country according to your index.

STAGE 3: What are the differences in economic development between UK regions?

In a 1996 United Nations list of the top 20 MEDCs, the UK was ranked 16th. This position is based on national average data. It does not reveal regional differences in levels of development. In the resources that follow you are provided with information on regional differences. You will see that differences in the level of economic development are matched by differences in the quality of life.

The reasons for the regional differences are largely connected to the growth and decline of particular industries. This topic is dealt with in more detail in the Enquiry on Where are the New Industries? (pages 196-215).

You are asked to use the information in the resources to compare the position in two contrasting regions. Draw up your findings in the form of a table with two columns (one for each region) and list the indicators (such as per capita GNP) down the left side. Briefly summarise what your table shows - is there a clear difference between your chosen regions? Do these statistics tell the full story?

Also in the resources you will find profiles of five UK families who are in different income groups. They all live in the same urban area. They are fairly typical of the area in which they live. Summarise what the information shows in the form of an article for a local newspaper.

Selected regional statistics for the UK

Region	GNP per capita (Index UK = 100)	Unemployment (% of work force)	Cars per 1000 people	16-18 year olds in FT education(%)
	1994	1996	1994	1995
South East	117	7.1	386	85
East Anglia	101	5.6	410	80
South West	95	6.1	410	82
West Midlands	93	7.6	392	78
East Midlands	96	6.8	350	76
Yorkshire and Humberside	89	8.0	326	76
North West	90	8.2	340	75
North	89	9.5	300	71
Wales	85	8.1	336	81
Scotland	100	7.9	303	87
Northern Ireland	82	11.6	326	79

The UK Index is calculated by working out the average GNP per capita for the whole country and then letting it equal 100. This allows an easy comparison to be made between regions. For example, the Index for the South East is 117, meaning that the region has a per capita GNP 17% higher than the national average. On the other hand, the North - with an Index of 89 - has a per capita GNP 11% lower than the national average.

Map of UK regions

Five typical families living in a major UK urban area

These profiles are based on the Rowntree Report into Income and Wealth published in 1995. The Report divided the population into fifths. Each family is a typical example from each of these fifths.

The poorest fifth (ie, 20%) of the population earns 8% of the total incomes earned in the UK.

Family A: Single mother with daughter, 8; living in a council flat near the city centre. She is unemployed and receives a variety of state benefits. Net income: £89 per week.

The next fifth earns 12% of total incomes.

Family B: Pensioner living on a council estate at the edge of the city. In addition to the Old Age Pension she receives a private occupational pension. Net income: £105 per week.

The middle fifth earns 16% of total incomes.

Family C: Couple aged 36 and 41 with son aged 11 and daughter aged 8; living in a new private house built on the city outskirts. Both work full time and receive Child Benefit. Net income: £289 per week.

The next fifth earns 23% of total incomes.

Family D: Couple aged 26 and 28 with no children, both working full time; living in a private 1930s detached house in the suburbs. Net income: £339 per week.

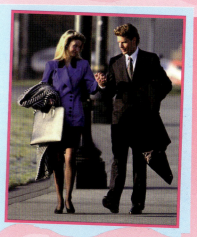

The richest fifth of the population earns 41% of total incomes.

Family E: Couple aged 58 and 57, 'empty nesters' - their grown up children have left home. Both work full time. They live in a small village 10 miles beyond the city boundary. Net income: £455 per week.

Is there a North-South Divide?

Although the national statistics show that, to some extent, the South of England is more prosperous than other regions, this does not tell the full story. There also exists a 'prosperity gap' between the inner cities and the suburbs (and rural areas). In general, average incomes and standards of living are lower in inner cities than in surrounding areas. So, there are areas of poverty in inner London and areas of prosperity in rural parts of Northern England.

This is illustrated by the map on page 30 which shows the location of the most and least deprived areas in Greater London. Because the South has few large cities, apart from London, and the North and West of Britain contains a large number (Cardiff, Belfast, Glasgow, Newcastle, Leeds, Sheffield, Liverpool and Manchester), it could be that the North-South Divide is just a reflection of the inner city / suburbia and rural area divide.

STAGE 4: Tanzania - a less economically developed country (LEDC)

Tanzania is an East African country approximately 4 times bigger in area than the UK. It has a population of almost 29 million (1996). In terms of its per capita GNP it is one of the poorest and least developed countries in the world. However, in Stage 1, you saw that it does better in terms of some indicators than the GNP data would suggest.

One of the problems that Tanzania faces is that its economy has failed to expand. In particular, the country has massive debts that it cannot repay. The interest on the loans and repayments are a huge burden on the economy which it can barely afford.

The resources that follow include information on the level of development in Tanzania. There is also an extract which outlines the serious debt burden that Tanzania faces.

Using the information provided, and working in pairs or in a small group, draw a spider (or star) diagram to illustrate Tanzania's current position. Compare your efforts with those of others.

List some of the reasons why Tanzania can be classed as an LEDC. Finally, write a brief note that explains how Tanzania's debt repayments are affecting the country's development.

Population
29 million, over 120 tribal groups - none of which make up more than 10% of the total
Life expectancy: 51 years
Infant mortality: 97 per 1,000 live births
Annual population growth rate: 3%
Rural population: 77% of total
Most Tanzanians are subsistence farmers. This means that they grow just enough food to feed themselves. Many are nomadic cattle herders - such as the Maasai. There is an Arab and Asian business community in the coastal cities.

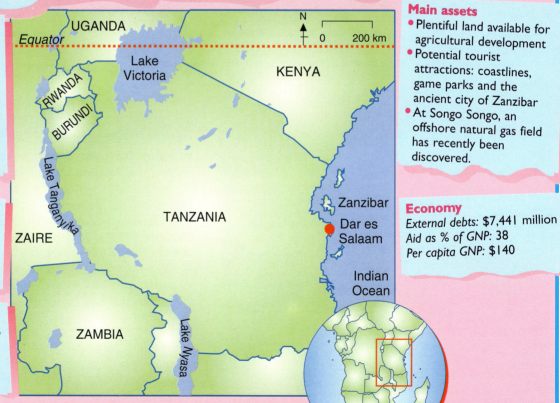

General
Area: 945,000 sq km (UK is 245,000 sq km)
Capital: Dar es Salaam
Access to safe water: 62% urban population; 46% rural population
Population with access to health care (within 5 km of their home): 76%.

Main exports
Coffee, cotton, (tourism is the third most important source of foreign income), tea, cashew nuts.

Climate
Tropical grassland (savanna), very hot wet season, hot dry season. Inland, highland areas are drier and cooler.

Main assets
• Plentiful land available for agricultural development
• Potential tourist attractions: coastlines, game parks and the ancient city of Zanzibar
• At Songo Songo, an offshore natural gas field has recently been discovered.

Economy
External debts: $7,441 million
Aid as % of GNP: 38
Per capita GNP: $140

Although Tanzania is rich in natural resources, its per capita GNP is low because, in the past:

- there has been low investment (ie, spending) on growth producing industries such as agriculture, manufacturing and power
- some investment has been wasted on unproductive schemes
- government policies of keeping food prices low have discouraged farmers from increasing their output
- before independence, the country was a British colony and its economy was run in Britain's interest, not Tanzania's
- the export crops that are produced have suffered from low and fluctuating prices on the world market
- foreign multinational companies have not been encouraged to set up operations in the country
- many people in the country are at a **subsistence level of income**, in other words they are too poor to save and invest in the future - they need all the money they can get just to stay alive
- population growth is so rapid that any increase in economic output is more than matched by the rise in the number of mouths to feed.

Tanzania's Debt Burden

Tanzania owes more than $7 billion in debts. This is money borrowed in the past to pay for development schemes. The interest and repayments on these debts were $155 million in 1994. This was more than double the amount that the government spent on public health care, including the provision of safe water. Yet more than 14 million people lack access to safe water, exposing them to the threat of water borne diseases (cholera, typhoid and dysentery), which are the major cause of premature death and disability.

Oxfam has highlighted the plight of villagers in the Arusha area of Tanzania. It is not uncommon for women to walk more than one hour each day to fetch drinking water from the nearest, often contaminated, well. This puts a huge physical burden on the women who could otherwise spend their time farming or looking after their children.

According to the World Bank, spending on health care would have to triple to provide a complete primary health care system. Although this is a massive sum, it is only slightly more than is now being spent on debt.

Education is another vital area that is being neglected. Primary education and adult literacy must be improved if the country is to break out of its cycle of poverty and deprivation. An Oxfam staff member, visiting a school in the Shinyanga area described the situation at the local primary school. 'There are two classes, with 50 children in each. One class has 5-8 year olds, the other has children aged 8-10. There are no books, desks or chairs. Children sit on the ground or on small rocks. They share the two or three pencils per class'.

Overall enrolment in primary schools has fallen from 90 per cent in 1980 to 70 per cent today. Literacy rates have halved in less than a decade. This is a picture of a country going backwards in development terms. The only solution in the short term is for the banks and governments in richer countries to agree to 'let off' Tanzania from its debts. This does not seem likely at the present time.

Aid is the term used for money, goods and human resources provided to LEDCs by richer countries. Most people believe that the main aim of aid should be to reduce poverty. However, there is a widespread view that aid does not always reach those with the greatest need.

Sometimes aid is given on the condition that it is used to buy goods from the donor country - which really the poor country does not want. Other times the aid is siphoned off to corrupt government officials who use it for themselves, or is spent by the military on armaments. Aid might be in the form of a grant but often it is a loan which must be repaid with interest.

Using the information that follows, briefly describe different forms of aid and give examples. What types of aid, in your view, are most helpful?

The photograph and material on Igombe Village contains information and questions in the form of a Development Compass Rose. After reading through all the resources, describe the issues faced by the villagers and answer the questions that are set around the photograph.

Some examples of aid

Type	Action	Example
ODA	Official Development Assistance (Includes bilateral and multilateral aid)	Food sent by the UK government to the Sudan
Bilateral Aid	Direct government to government aid	UK government aid to India, eg funds to improve a State Electricity Board.
Multilateral Aid	Aid that is channelled between countries through an international body such as the United Nations	World Bank, World Health Organisation (WHO) also channel multilateral aid
NGO Aid	Non government organisations (usually charities) distribute money and expertise from MEDCs to LEDCs	Oxfam, Action Aid, Help the Aged, Medecin sans Frontieres

Aid can take many forms. It can range from emergency food relief to massive dam building projects. In Tanzania, some international aid has been used to develop tourism in the country's wildlife reserves.

Aid statistics

In 1994, total world ODA was $59 billion. 70% was bilateral, government to government, aid and 30% was multilateral aid. The total aid from NGOs was 10% of 'official' aid (ie, $5.9 billion).

The UN target is for More Economically Developed Countries to give 0.7% of their GNP as aid. Only Norway, Denmark, Sweden and the Netherlands reach this target. The UK currently gives about 0.3% of its GNP in aid. On a world scale, total official aid is at its lowest point for 20 years.

Although they do not reach the UN target, Japan and the USA are the world's biggest aid donors because their GNP is so large.
In 1994, Japan gave $13 billion and the USA gave $10 billion.

International aid is a 'drop in the ocean' compared with the amount needed to raise living standards to Western levels.

Aid has not managed to remove poverty. About 1.5 billion people live in 100 countries that are poorer now than they used to be. Some, including Ghana, Haiti, Liberia, Nicaragua, Rwanda, Sudan, Venezuela, Zaire and Zambia, were richer in 1960 than they are now. Most of these countries are major recipients of aid.

Over the past decade the pattern of aid has shifted. Previously, most was given for spending on health and education, now most is given for emergencies and debt relief. This latter category is not 'new' money, it is simply the richer countries 'letting off' the poorer countries from repaying part of their debts.

The main aid recipients are China ($3bn), Egypt ($2.3 bn), Indonesia ($2.0 bn) and India ($1.5 bn).

ActionAid - an NGO

ActionAid exists to help overcome poverty and improve the quality of life of people in the developing world. The organisation works directly with three million of the poorest children, families and communities in 20 countries throughout Africa, Asia and Latin America. It employs 3,000 staff, mainly nationals in their own countries who know the needs of local communities.

In 1995 ActionAid raised over £37 million. Funds mainly come from private donors, other funds come from companies and the British Government.
Current projects include the provision of clean and safe water supplies as well as education, especially adult literacy and primary education. In 1995, 13% of Action Aid's money went on emergency relief - mostly to refugees from wars and people affected by drought.

Igombe Village - Tanzania

Igombe is a small farming and fishing community on the shore of Lake Victoria. By tradition, most men in the village are fishermen, most women are farmers.

Fish are caught using nets which are set out in the lake each day. In the morning, the men paddle out in canoes - taking an hour or more - to check their nets. An average catch is 50 fish, each weighing between 0.5 and 2.5 kilos. On returning to the village the men keep some fish for their own families and sell the rest to a dealer. For this they receive between 20 and 30 pence per kilo depending on market prices. The dealer transports and sells the fish to a wholesaler in a nearby town. The wholesaler then fillets and freezes the fish which are sold in the capital, Dar es Salaam and are exported to Kenya. The fishermen only receive a small fraction of the price at which the fish are finally sold.

The women of the family spend most of their time tending small plots of land on which they grow maize, cassava, tomatoes and onions. All the work is done by hand. The families' diet is of fish and the food they grow themselves. Their homes have little furniture apart from a table, chairs and beds. They have no electricity or running water. Most of the people in the village have a subsistence level of income. In other words, they have just enough to feed and clothe themselves. Money does not play a big part in people's lives because they produce so much of their own needs. The small amount they earn from fishing is used to buy the things that they cannot produce themselves.

There is a primary school in the village which 5 to 8 year olds attend. After that age, most youngsters help with the family tasks, either helping with the fishing, hoeing and planting crops, or carrying out household chores. A mobile clinic visits the area once a month. However, secondary education and more advanced medical care is only available two hours away by bus. This is a journey that few families can afford to make on a regular basis.

There are no social security or pension schemes for the villagers. When they become ill or too old to look after themselves, they must rely on their family for help. Local money lenders charge very high rates of interest and the nearest bank and government office are two hours away.

A family in Igombe mending their fishing nets.

Development Compass

Natural

"" **How might the natural resources of the lake be better used?** ""

"" **How might the villagers increase their incomes by using the land differently?** ""

"" **How might the men receive more income from their fishing?** ""

"" **Who decides the price of the fish?** ""

Who decides?

"" **What type of aid would most help the villagers?** ""

Economic

"" **What part do women play in the local economy?** ""

"" **Why is it in the economic interests of the families to have large numbers of children?** ""

"" **Where might the villagers get loans to improve the equipment they use?** ""

"" **Does local tradition and culture help or hinder progress?** ""

"" **Do the villagers have a balanced diet?** ""

"" **What, in your view, is the standard of living of the villagers?** ""

Social

"" **What do the villagers need to improve their quality of life?** ""

Note: a development rose helps focus attention on the key issues:
Natural environment - how does it affect people?
Who makes the political and other decisions and who has power over people's lives?
Economic factors - what are they?
Social structures, culture and tradition - what influence do they have?

Aid in Tanzania

Aid comes in a number of forms. Emergency relief is designed to give immediate help to people who are in danger of starving. In northern Tanzania, tens of thousands of Rwandan refugees were fed, clothed and housed by relief agencies in 1995 and 1996. This type of aid keeps people alive but does not create long term prosperity.

Some bilateral and multilateral aid goes to the Tanzanian government. This is used to build roads, power lines and to improve port facilities. Unfortunately it is not always used wisely and, if it does not bring sufficient economic return, it simply increases the country's debt without raising incomes and the tax revenues which are needed to repay the loan.

In many people's view, the best aid comes from NGOs and small scale government schemes financed by foreign aid. When local people are involved, especially in self help schemes, there is less chance that the money will go 'astray' or be wasted.

Often, the amount of money needed to start self help schemes is surprisingly small. In one village in southern Tanzania a group of women borrowed £2,000 from the government (under a UNICEF funded scheme) to set up a small flour mill. The money was used to buy building materials and the milling equipment. The women supply the labour for the mill.

This type of venture, based on a knowledge of local conditions, is sometimes called 'appropriate technology'. It is often more successful than larger scale schemes using foreign, high technology. The women's cooperative has used its revenue from selling flour to reinvest in other farming schemes. The villagers now grow spinach, aubergines and onions to eat and for sale at the local market. With their spare income, they have started to re-equip the junior school and to improve their own standard of living.

Aid agencies and foreign governments are slowly learning from the mistakes made in previous decades. One such example was an ultra modern cotton textile factory built with foreign aid in the capital in the 1970s. The machinery was so 'high-tech' that local people could not operate or repair it. Because it was very automated, few jobs were created. The factory needed the highest quality raw cotton which was not available in Tanzania, so had to be imported. Finally, the expensive cotton fabric produced at the mill cost too much for most local people to buy.

The whole scheme was an expensive disaster. It was a waste of money and it did not help the country or its people.

These women are using 'appropriate technology' to raise their own living standards.

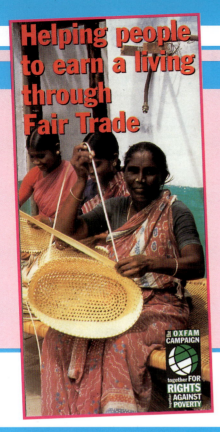

Helping people to earn a living through Fair Trade

Oxfam Fair Trade Campaign

Oxfam is well known for its emergency relief and long term development work in poor countries. In recent years it has begun another type of aid. Believing that one of the best ways to help poor people is to encourage self help, Oxfam is campaigning to provide people in less developed countries with a decent wage.

In some cases, Oxfam and other agencies like Traidcraft buy direct from suppliers in the Third World and sell the goods for 'fair' prices in Britain and elsewhere. These goods might be food, coffee, textiles or handicrafts.

At the same time, Oxfam is campaigning for fair wages to be paid in the factories of less developed countries that supply the developed world. In particular, in 1996, it highlighted the low wages paid in South East Asia to people making trainers and textile goods.

STAGE 6: Review

As a means of reviewing and summarising your work on Development, you are asked to take the role of an employee of an aid agency such as Oxfam. Your task is to write a public information leaflet that sets out the answers to the following questions:

- What is 'development' and how can it be measured?
- Where are the most and least economically developed countries?
- Why is development considered important?
- What are the problems faced by LEDCs and how can they be overcome?

Illustrate your leaflet with statistics, photographs or maps which help explain your points. Try and obtain actual examples to look at before you start.

Glossary

All the terms listed below are explained in this Enquiry. In each case, write a definition of the term and, if possible, give at least one example.

Development	**LEDC**
Development indicator	**Consumer goods**
Economic development	**Services**
Social development	**Human Development Index (HDI)**
Gross National Product (GNP)	**Aid**
Per capita Gross National Product	**Subsistence level of income**
MEDC	**Fair Trade**

(Note that the term Gross Domestic Product (GDP) is sometimes used instead of GNP. You can use either GNP or GDP as a measure of a country's National Income.)

6 Coping with the weather

to the student

Weather affects all our lives. At the simplest level, it influences the clothes we wear and whether we switch on the heating. In the United Kingdom a spell of dry weather is enjoyed by most people. However, it can cause problems if it persists and becomes a drought. For example, a hose pipe ban might be imposed preventing people from watering their gardens or washing their cars. Farmers might find that their crops fail to grow and this might cause the price of, say, potatoes to rise if there is a shortage. In less economically developed countries, a drought can cause much more severe problems such as hunger and even starvation.

In these Enquiries, we consider the factors that affect the weather and compare the UK's weather with other parts of the world. In addition, we shall see how the weather affects people's lives and also consider how people cope with extremes of weather such as hurricanes and droughts.

questions to consider

1 What factors influence the weather?
2 How does the weather in the United Kingdom compare with other regions of the world?
3 How does the weather affect human activity?
4 How do people cope with, and respond to, extremes of weather such as hurricanes and drought?

key ideas

Weather is the atmospheric condition at any one time. It includes the precipitation, temperature, wind speed and direction, sunshine, humidity and air pressure.

Climate is a term that describes the average weather conditions in a particular place over a period of time.

Precipitation is the collective term for all forms of moisture that falls to the ground such as rain, snow, sleet and hail.

activities

Using information from this and the facing page:

Working with a partner,
a) List factors that influence the weather.
b) List ways in which weather can affect human activity.

FACTORS THAT INFLUENCE THE WEATHER

Latitude, time of day and time of year

Altitude and aspect

Prevailing wind

Distance from the sea

FOG

SLEET

SNOW

HURRICANE

MIST

WIND

THUNDERSTORM

SHOWERS

LIGHTNING

RAIN

HEATWAVE

Which crops to grow

Where to go on holiday

Where to set up solar panels

Coping with emergencies

HOW DOES THE WEATHER AFFECT PEOPLE'S LIVES

nquiry

UK Weather

In this Enquiry you will study the factors that influence the weather. You will also investigate different patterns of weather and consider how these affect human activity. The outcome of the Enquiry will be a written account of your local weather. This will include the factors that influence your weather and a summary of how the weather affects people's lives.

STAGE 1: What factors influence the weather?

The resources that follow outline the main factors that influence weather. Remember that 'weather' is a term that describes the day to day atmospheric conditions in a particular location. For example, a weather report on the 1st October for Bournemouth read:

Sunshine (hours)	Rain (mm)	Temperature (degrees Celsius)	
		High	Low
8.8	2	17	8

A more detailed report would also include wind speed and direction, humidity and atmospheric pressure. Over a number of years, an average of similar readings could be calculated and used to describe the climate of Bournemouth. Using the resources that follow, make notes in your own words on the factors that influence weather. Write a paragraph on each using the following sub headings:

Latitude
Time of year
Height above sea level (altitude)
Aspect (ie, direction the land is facing)

Nearness to the sea
Prevailing wind direction
Types of rainfall
Air mass

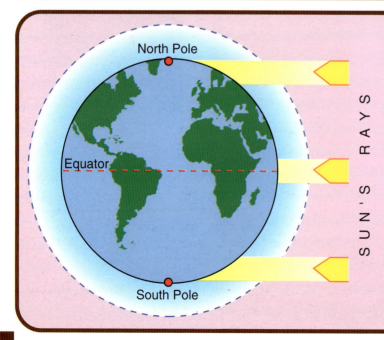

Latitude

Lines of latitude are parallel to the Equator. London is on the line of latitude 51 degrees North of the Equator. The North Pole is 90 degrees North.
At higher latitudes, the Sun's rays do not heat the Earth's surface as much as they do at the Equator. This is for two reasons:
- the curvature of the Earth causes the Sun's rays to be angled directly downward at midday near the Equator. Closer to the Poles, the Sun's rays shine at a lower angle. So, near the Equator, the Sun's energy is concentrated upon a much smaller area than near the Poles.
- at the Equator, the Sun's rays have to pass through less atmosphere than nearer the Poles. Because the atmosphere contains water and dust particles that absorb energy, at the Poles more heat is lost before the Sun's rays reach the ground.

Time of year / season

The Earth orbits the Sun once a year. It lies at an angle or tilt in relation to the Sun, as shown on the diagram. At the same time, the Earth revolves on its own axis once every 24 hours.

In mid-December the angle of tilt causes the South Pole to have 24 hours of sunlight every day and the North Pole to have 24 hours of darkness. Six months later the position is reversed.

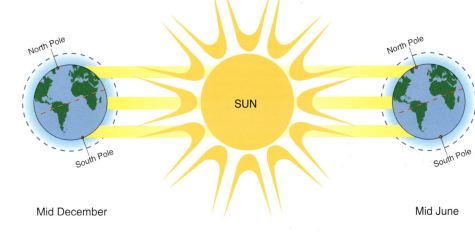

Mid December   Mid June

From the diagram you can see that in December, the Sun's rays are at a much lower angle in the northern hemisphere than in the south. In June, the opposite is true and it is summer in the northern hemisphere.

So, there are two reasons for our seasons:
- the Earth orbits the Sun once a year
- the Earth's axis is tilted causing the Sun to be more directly overhead in our summer compared with our winter.

Height above sea level (altitude) and aspect

The atmosphere is heated not directly by sunlight but by warmth radiated from the Earth's surface. So, the higher the altitude, the cooler it becomes. On average, temperature falls by 6° Celsius per 1000 metres.

Temperature falls as height increases. When the temperature falls below freezing, there is a snowline.

Slopes that face the Sun (ie, face south in the northern hemisphere) are warmer than those in shade because they receive more solar radiation (ie, sunshine).

Prevailing wind and nearness to the ocean

Land heats up and cools down more quickly than seawater. So, in summer the continental land mass heats up faster than the Atlantic Ocean. Air temperatures above the land are higher than those over the ocean.

In winter the opposite is true. The land cools faster than the sea so the continental landmass is much colder than the Atlantic Ocean.

[On a more local scale, inland Britain is warmer in summer and cooler in winter than coastal areas.]

So, an important factor affecting temperature is nearness to the sea or ocean. Another factor is the direction of the wind. If wind blows from the east, Britain gets weather typical of the continental landmass - very cold in winter, hot in summer. However, most of the time, Britain gets wind from the south west. This is the **prevailing** wind. It brings Atlantic weather, relatively mild in Winter, cool in Summer.

What causes rainfall?

Air contains water vapour. This is a gas but it turns into water droplets when the temperature falls. The process is called **condensation**. When sufficient water vapour condenses, clouds form and rain can fall. There are 3 main atmospheric processes that cause air to cool, condense and form clouds. These give rise to 3 types of rainfall:
- relief rainfall
- convection rainfall
- frontal rainfall.

Relief rainfall

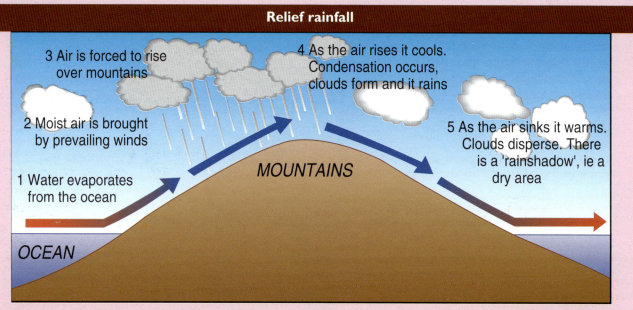

3 Air is forced to rise over mountains

4 As the air rises it cools. Condensation occurs, clouds form and it rains

2 Moist air is brought by prevailing winds

5 As the air sinks it warms. Clouds disperse. There is a 'rainshadow', ie a dry area

1 Water evaporates from the ocean

MOUNTAINS

OCEAN

Relief is simply another name for the shape of the landscape. In this context it is used to describe hills and mountains. Air is cooled as it is forced to rise over high ground.

Convection rainfall

The Sun's rays warm the ground and sea. These then heat the air which starts to rise in convection currents. As the air rises, it cools and water condenses. Clouds form and rain can occur. When the updraught is rapid the air can rise to 30,000 feet and thunderstorms can occur.

Frontal rainfall

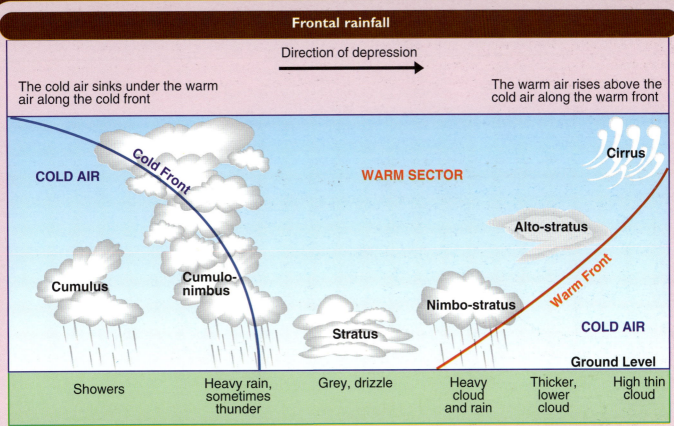

Direction of depression →

The cold air sinks under the warm air along the cold front

The warm air rises above the cold air along the warm front

COLD AIR
Cold Front
WARM SECTOR
Cirrus
Alto-stratus
Cumulus
Cumulo-nimbus
Nimbo-stratus
Stratus
Warm Front
COLD AIR
Ground Level

| Showers | Heavy rain, sometimes thunder | Grey, drizzle | Heavy cloud and rain | Thicker, lower cloud | High thin cloud |

Swirling masses of rising air often form over oceans. Because the air is rising, there is **low atmospheric pressure**. Frontal systems often form where warm and cold air masses meet. They are called **depressions**. The boundary where the cold and warm air meets is called a **front**. As the warm air rises above the cold air, it cools. Water vapour condenses and clouds form.

Air masses that affect UK weather

One of the reasons why the weather is so variable in the UK is because depressions and anti-cyclones 'suck' in air from different directions as they pass over.

Air masses that cross the Atlantic - called **Maritime** - tend to be moist.
Air masses from Europe and Asia - called **Continental** - tend to be dry.

Air masses from the south - called **Tropical** - tend to be warm.
Air masses from the north - called **Polar** or **Arctic** - tend to be cold or very cold.

Note: swirling masses of sinking air are called **anticyclones**. In these systems there is **high atmospheric pressure**. They bring dry, settled weather. In the UK they cause hot, sunny weather in summer. In winter they can also bring clear, sunny weather but sometimes they also cause grey, foggy weather.

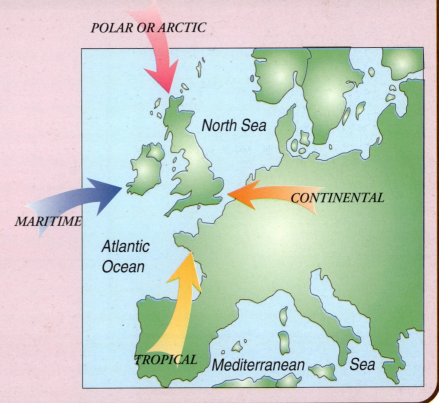

POLAR OR ARCTIC

North Sea

CONTINENTAL

MARITIME

Atlantic Ocean

TROPICAL Mediterranean Sea

The three main factors that influence the UK's weather are:

- The United Kingdom is situated in temperate latitudes (in between Tropical and Arctic latitudes).
- It is situated on the western margin of the European land mass, facing the Atlantic Ocean.
- The prevailing wind is westerly. In other words, the wind direction that is most common is from the west, blowing across the Atlantic.

In broad terms, the UK's climate (ie, average weather conditions) can be described as **temperate**:
- warm, wet summers with cool, wet winters.

Average January temperatures are generally between 3 and 5°C. Average July temperatures are generally between 14 and 17°C. Average annual rainfall is over 2000 mm in the north west and below 500 mm in the south east. Most of the country has between 600 and 1200 mm per year.

There is a small average **temperature range** (this is the difference between the highest and lowest) and it is wet throughout the year. However, these are the average conditions and the UK's weather is particularly variable. This means that it varies from the average a great deal, even from day to day. There can be warm days in winter and cool days in summer. Likewise, at any time of the year, there can be long dry spells.

As well as having variable weather from day to day, there are important differences in the weather pattern between parts of the UK. In general terms, it is wetter in the west and drier in the east. In summer it is warmest in the south, however in winter it is warmest in the west. The resources that follow contain more detailed information on the UK's weather pattern.

After you have read through the resources, make notes under the headings:
a) What is the pattern of weather in the UK?
b) Why is the UK's weather so variable?

On an outline map of the UK, mark on the summer and winter isotherms, and the wettest and driest areas. Write short captions on the map that explain why the differences occur. In other words, why is it wettest in the west, why is it warmest in the south in summer?

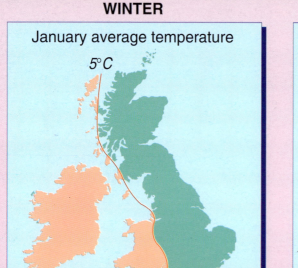

Temperature

WINTER

January average temperature

SUMMER

July average temperature

The maps show two **isotherms**. These are lines that join places which have the same temperature.
In winter there is an east-west divide in average temperatures. The west, and coastal areas, are mildest.
In summer there is a north-south divide. The south, and inland areas, are warmest.

Rainfall

The north and west of the British Isles receive the highest rainfall. This is because they are more mountainous (relief rainfall) and are nearest the Atlantic Ocean, over which the prevailing wind blows (frontal rainfall).

The south and east are drier and warmer. They receive less relief and frontal rain than the north and west, but do get about half their total rain in summer from convectional rain, sometimes accompanied by thunderstorms.

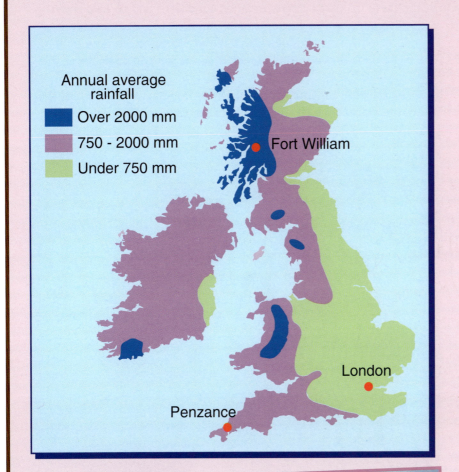

Annual average rainfall
- Over 2000 mm
- 750 – 2000 mm
- Under 750 mm

Fort William

London

Penzance

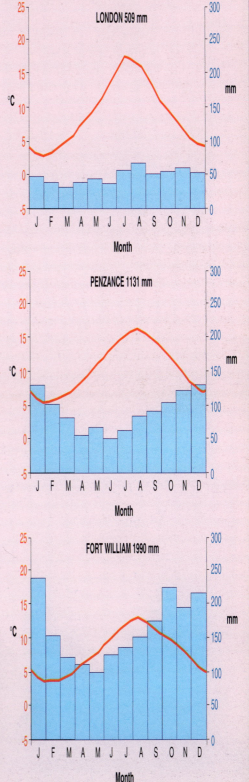

LONDON 509 mm

PENZANCE 1131 mm

FORT WILLIAM 1990 mm

Highest rainfall occurs on mountains near the west coast of the British Isles.

Forecasting the weather

Britain's variable weather is mainly due to the passage of depressions. As the fronts pass over they bring rain, with drier conditions in between. Within the depressions there is warm air and colder air. Because the air circulates (in an anti-clockwise direction), as the depression passes over, the wind direction changes. So, for example, the wind might start from a southerly direction bringing mild air, but then move to the north bringing colder air.

The Meteorological Office produces 'synoptic charts' to plot current weather conditions and to help them make forecasts. They use satellites together with observers on the ground and on ships to monitor pressure, temperature and other climatic conditions. The chart below was produced by the Met Office at the time of an exceptionally severe storm in January 1993. The high winds drove the oil tanker Brear ashore on to rocks at the southern tip of the Shetland Isles.

The tightly packed isobars (lines) on the chart indicate very low pressure and strong winds at the centre of the depression. To the south, near the Mediterranean, the isobars are far apart - indicating calm and settled weather.

Storm force winds can bring danger to shipping. The Met Office broadcasts shipping forecasts to warn sailors of likely conditions.

The Met Office issues severe weather warnings if there is a likelihood of heavy snow, flooding, strong winds or fog.

Satellite photograph and weather chart for the British Isles (September 1993)

The weather chart (or synoptic chart) with its symbols shows a frontal system approaching the British Isles. The satellite photograph was taken at the same time.

Most of Britain is under clear skies with small cumulus clouds. High cirrus clouds over the Irish Sea are signs of the approaching front. Spain and the western Mediterranean were enjoying hot, settled weather under the high pressure anticyclone.

As the front approached Ireland thicker clouds brought rain. Behind the frontal system, a cool northerly air stream was bringing showery rain.

The area of low pressure at the centre of the depression is clearly seen as a swirl of clouds. The cold front had caught up with the warm front (forming an occluded front) along the northern part of the frontal system.

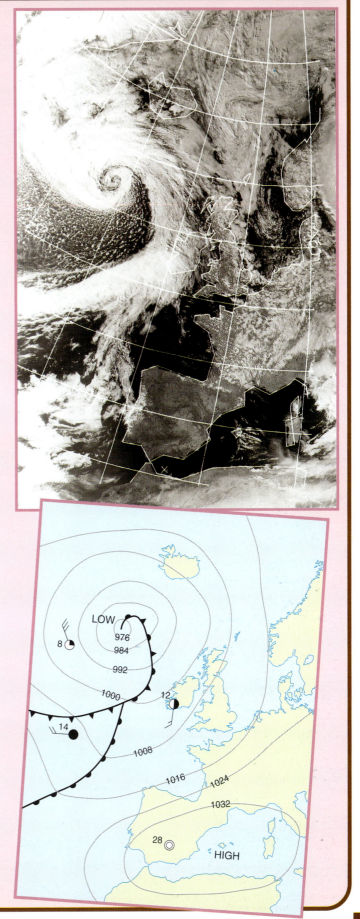

- **976 -** Isobars are drawn at intervals of eight millibars

Temperature is given in degrees C and is shown on the charts by means of figures alongside the station circle

Cloud cover (oktas)

○ 0	◑ 3	◕ 6	⊗ Sky obscured
◔ 1 or less	◐ 4	⬤ 7	⊗ Missing or doubtful data
◕ 2	◔ 5	⬤ 8	

Wind speed (knots) and **Direction**

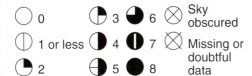

◎ Calm   ○⎯ 1-2

○⎯ 8-12   For each additional half feather add 5 knots

Precipitation

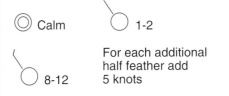

● Rain	⸼ Drizzle	✳ Snow
● Rain and ✳ snow	▽ Shower	△ Hail
⚡ Thunderstorm	═ Mist	≡ Fog

Fronts

Warm front at the surface   Cold front at the surface   Occluded front (cold front has caught warm front)

STAGE 3: Climates in other parts of the world

There are many different climatic zones in the world. They include:

- equatorial rainforests which are hot and wet, eg Amazonia
- deserts which are hot and dry, eg Sahara
- Mediterranean regions which have hot and dry summers, mild and wet winters

Using the following resources, and the information from Stage 2, compare the climate in two different regions with that in the UK. Make your notes in the form of a table, for example:

	UK	Amazonia	Sahara	Mediterranean
Average temperature	Winter 3-5°C Summer 14-17°C			
Average precipitation	600-1200 mm per year Rain all year round			
Variability	Very variable day to day but few extremes			

Mark on a world map the climate zones you have described.

Three climatic zones

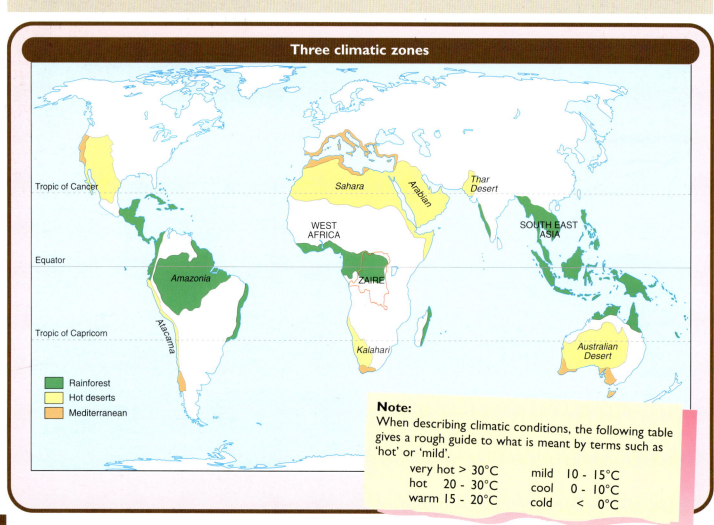

Tropic of Cancer

Sahara Arabian Thar Desert

WEST AFRICA SOUTH EAST ASIA

Equator

Amazonia ZAIRE

Tropic of Capricorn

Atacama Kalahari Australian Desert

- Rainforest
- Hot deserts
- Mediterranean

Note:
When describing climatic conditions, the following table gives a rough guide to what is meant by terms such as 'hot' or 'mild'.

very hot	> 30°C	mild	10 - 15°C
hot	20 - 30°C	cool	0 - 10°C
warm	15 - 20°C	cold	< 0°C

Amazonia - an Equatorial rainforest climate

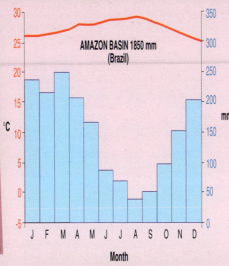

AMAZON BASIN 1850 mm
(Brazil)

The climate can be described as hot and wet throughout the year. There is no summer or winter, the temperature range is very small. However, between June and October there is a drier season. For the rest of the year, heavy convectional thunderstorms occur on most days. The weather pattern tends to be very predictable.

Sahara - a hot desert climate

The climate is hot and dry throughout the year. In early summer it is very hot, sometimes over 40°C at midday.

There is a chance of heavy rainstorms in July and August but these are infrequent and unreliable. Most of the time, the weather is settled and predictable. There is a big temperature range between January and July, and also between night and day. Without any cloud cover to retain the heat, the desert temperature can fall below freezing at night.

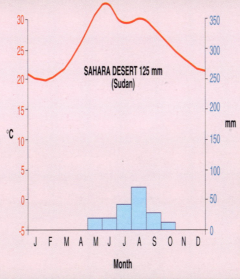

SAHARA DESERT 125 mm
(Sudan)

Greece - a Mediterranean climate

The climate is hot and dry in summer, mild and wet in winter. There is a big annual temperature range. Winter rains come mainly in the form of frontal depressions. Summer rain usually comes in the form of severe thunderstorms. However, for most of the time, summer weather tends to be dry and settled. In winter, the weather is more like that in Britain, only warmer.

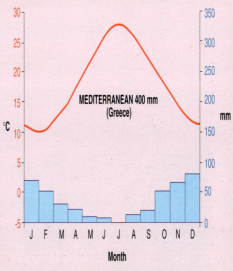

MEDITERRANEAN 400 mm
(Greece)

Human activity is a term that covers many things, such as:

- farming
- communication
- settlement
- holidays and recreation.

Weather can have an important impact on all of these (as well as your decision on what clothes to wear). In the resources that follow, there are examples of how people are affected by the weather. Using the information, list a number of human activities and outline how they are affected by the weather. Discuss your list with a partner and try to add extra examples from your own knowledge and experience. Then write a more detailed account that describes and explains one of the examples on your list.

An Alpine Valley in Northern Italy

In Alpine regions, the landscape causes differences in the **microclimate** of the valley sides and valley bottom. In other words, there are different climatic conditions depending on the altitude and aspect of the valley slopes. A microclimate is simply the climate of a small local area.

On north facing slopes, conditions are colder and wetter. These conditions are less suited to farming so the land is often left uncleared as forest. There are few houses.

The south facing slopes receive longer and more direct sunlight than north facing slopes. This causes them to be warmer and drier (because evaporation is higher).

On south facing slopes, most farming occurs; farmers grow grass for hay and keep sheep, goats and cattle on rich pastures. Farmhouses and villages are built on this warmer side of the valley.

On flat, lower valley bottoms, the alluvial soil is fertile. However, crops which can be damaged by frost - such as vines and fruit - are not grown on the bottom land. This is because cold air sinks during the night and creates 'frost pockets' on the lowest levels.

'Motorists face flood chaos on roads'

Motorists in the West Country were advised last night to beware of flooded roads. Torrential rain had caused rivers to overflow and threatened several major roads.

- *newspaper report, March 6th 1996.*

A cancer that threatens sun seekers

Britain is near the top of the European league table for cases of malignant melanoma - skin cancer. While dark skinned southern Europeans go about their normal business in the sun, taking shade in the hottest part of the day and building a gradual tan, fair skinned Britons behave differently. The first sign of sun in Britain sees many office workers peeling off their clothing and sitting outside for lengthy periods. They do not try to tan gradually because it might rain the next week.

The result has been a doubling of skin cancer cases in a decade. Foreign holidays - a fortnight in the sun - have also played their part. In Victorian times the fashion was to shield the skin from the sun, because a tan was a sign of having a menial outdoor job. Perhaps the wheel is turning full circle, with health conscious people once again avoiding the sun and taking their holidays in places which are cooler and less sunny.

'Blizzards give ski resorts a lift'

Heavy snow has given Scottish ski resorts better conditions than some continental destinations. Four of Scotland's five resorts were open yesterday as blue skies and calm winds returned to the Highlands. Hotels are expecting a last minute rush of tourists over the holiday period.

The variable Scottish weather always makes skiing holidays there a riskier business than in the Alps. Cold weather, with a deep base of snow is vital for ski resorts. In Aviemore, just as in the Alps, many hundreds of jobs depend upon the weather. No snow means no winter skiers.

Ironically, if the snow is too deep and conditions are too icy, the roads become impassable and the tourists cannot get through. So, too much snow can cause as many problems as too little.

- *newspaper report, December 1995.*

STAGE 5: Review

In this Enquiry you have considered: **what** the weather is like in different places, **what** factors influence the weather and **how** weather affects people's lives.

To complete your Enquiry, you are now asked to write an account of the weather in your home region. Use the information provided together with information from your teacher. You might be able to research the topic using your local newspaper if it carries a daily weather report. Other sources are atlases, books about your region, the local water company and gardening clubs. You should describe your local climate (ie, the weather conditions over a whole year). Explain the main factors that influence your weather and describe how the weather affects the lives of local people.

Glossary

All the terms listed below are explained in this Enquiry. In each case, write a definition of the term and, if possible, give at least one example.

Weather	**Relief rainfall**	**Anticyclone**
Climate	**Relief**	**High pressure**
Precipitation	**Rain shadow**	**Temperate**
Latitude	**Convection rainfall**	**Temperature range**
Altitude	**Frontal rainfall**	**Isotherm**
Aspect	**Low pressure**	**Synoptic chart**
Prevailing wind	**Depression**	**Microclimate**
Condensation	**Front**	

enquiry

Severe weather

Hurricanes, floods and drought are extreme weather conditions that can devastate whole regions and disrupt people's lives. The worst disruption generally occurs in less economically developed countries. This is because they have fewer resources to plan and prepare for the extreme conditions.

An issue that causes great concern for the future is *global warming*. If the trend of rising average temperatures continues, there is a danger that polar ice caps will melt and raise the sea level. This will cause flooding in coastal regions. The rise in temperature will also have an unpredictable effect on world weather patterns. Some places will become wetter while others will become drier.

The Enquiry that follows describes where particular hurricanes and droughts have occurred and how people have responded. It also describes how governments and international bodies are responding to the threat of global warming. The outcome of the Enquiry will involve you taking the role of a government official in a less economically developed country of your choice. Your task is to write a draft plan of how your country could better cope with a particular extreme weather condition (eg, storm or drought). You will be asked to outline the problems that face your country and suggest possible solutions.

STAGE 1: Why do tropical storms and hurricanes occur?

The largest and most severe weather systems occur over tropical oceans in summer. In the Atlantic, we call them hurricanes. The information that follows describes where and how they occur.

On a flow diagram, make notes that explain how hurricanes start. On a world map, mark on the location of where hurricanes, typhoons and cyclones occur. Mark the path of Hurricane Cesar that is described in the resources.

The distribution of tropical storms and hurricanes

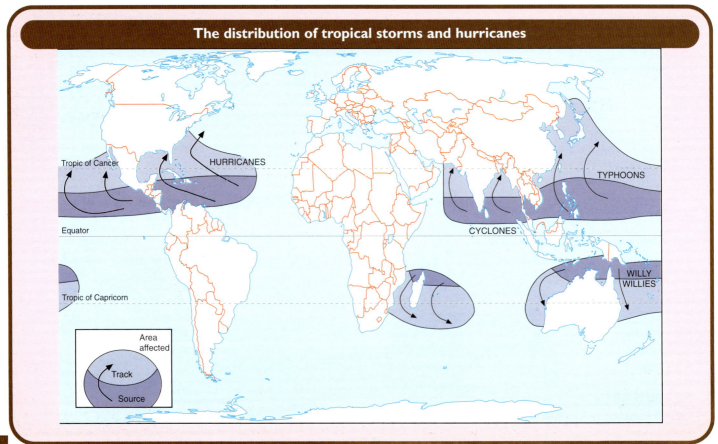

Hurricanes

In the Pacific they are called typhoons, in the Indian Ocean they are called cyclones. In the Atlantic they are known as hurricanes. They have different names but they all start in the same way.

As sea surface temperatures rise in the tropics in summer, massive quantities of water evaporate and form storm clouds. The Earth's rotation causes the storm clouds to spin (anti clockwise in the northern hemisphere, clockwise in the southern). Each day, as much as 200 million tons of seawater are recycled by the strengthening storm, firstly evaporating and then falling as rain.

When the winds exceed 75 miles per hour, the storm is classified as a hurricane. It can be up to 800 km in diameter. The path of hurricanes is erratic and unpredictable but their average speed is only 12 miles per hour. The Atlantic hurricane season runs from June 1st to November 30th. Each **tropical storm** (winds over 40 mph but less than a hurricane) and hurricane is given a name. The first of the season has a name beginning with A, and so on. Since 1979, hurricanes in the North Atlantic have been given male and female names alternatively. Names of particularly destructive hurricanes are not used twice.

A tropical storm - the start of a new hurricane

Powerful swirling winds and cloud patterns are the first sign of a new hurricane.

The growth of hurricanes

Stage 1

1 Tropical sun provides very high insolation levels during summer

4 Water rapidly evaporates forming water vapour-laden air

3 Air above sea heats up, and therefore rises (convection), causing very low pressure

2 Sea absorbs sun's energy gradually heating up and radiating heat to air

Stage 2

6 During condensation energy is released allowing further convection

7 Towering clouds result

5 As air rises it cools causing water vapour to condense creating water droplets

8 As water droplets combine they fall as intense rainfall

10 Air is sucked in towards the centre of the hurricane to replace convected air. This creates strong winds. The greater the uplift the stronger the winds

9 Sea continues to heat air, so processes of evaporation, convection and condensation are on-going

11 The earth's rotation causes the winds to spiral inwards

Stage 3

12 At about 10km above sea level air cools down enough to stop rising

15 Air spreads out from centre of hurricane at upper level, gradually sinking at edge of hurricane. This will be sucked back in towards the centre

14 Some air sinks in centre of hurricane - the eye, causing high pressure and calm conditions

13 Air cannot directly sink towards sea due to continued convection from below

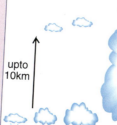

upto 10km

up to 400km radius

The information in the resources which follow describes how people in different parts of the world cope with tropical storms and hurricanes.

Make a brief set of notes which compare the impact of Hurricane Andrew in Florida with the two cyclones in the Bay of Bengal. Explain why there was a much greater loss of life in Bengal.

The resources also include a diary of a holidaymaker in Costa Rica when Hurricane Cesar hit. Describe the effects of the hurricane and make a list of things that the people and government could do to lessen the impact of future hurricanes.

Hurricane Andrew

In the United States, the National Hurricane Centre monitors all tropical storms in the North Atlantic. On August 17th 1992 the first tropical storm of the season was spotted off the coast of west Africa. It was named Andrew. The storm was tracked by satellite and by specially equipped aircraft as it built up to hurricane force. By Sunday 24th August hurricane Andrew was approaching the coast of South Florida. Warnings were broadcast on TV and radio for people to board up their homes and to evacuate the area. Because most families in Florida have cars, the highways were soon busy with people escaping.

The people who chose to stay were advised to sit in the most enclosed room of their house - often the bathroom - and cover themselves with a mattress. Hurricane Andrew turned out to be the costliest natural disaster ever in the USA. The strongest winds, in the 'eye wall' reached over 164 miles per hour - as strong as a tornado. One person living in the path of the storm reported: 'The front door exploded. I could hear the roof ripping off. When the eye passed over I could see starlight above'. Wooden houses and trailer parks were smashed, however most brick and concrete buildings survived.

Communications completely broke down as overhead lines were damaged. The National Guard was called out to maintain order and to help in restoring essential services. The final cost was 58 deaths and $25 billion in damages. It was noted that the worst damage occurred in 'streaks' only a few hundred yards long and 20 to 30 feet wide. This suggests highly localised and intense winds that only last for a few minutes, like a tornado, must occur in the severest hurricanes.

Hurricane Andrew with its eye clearly visible. It has passed over Florida and is in the Gulf of Mexico.

Hurricane damage.

Cyclones in the Bay of Bengal

In November 1970, a cyclone moving north in the Bay of Bengal caused the sea to rise 30 feet above its normal level. The storm hit the low lying Ganges delta and caused at least 300,000 deaths and wrecked 600,000 dwellings. The same thing happened in 1991. This time, 140,000 died and 1.4 million houses were destroyed.

Because Bangladesh is a poor country, it does not have the resources to plan and prepare for tropical storms in the same way that, for instance, Florida is able to do. The transport system is undeveloped and relatively few people have cars - making it difficult for people to flee inland. Sophisticated tracking and warning systems are not in place but, even if they were, there is little that most people can do. Houses are generally flimsy and made of wood, there are few concrete storm shelters. The land is low lying and only a few areas have sea walls that can withstand the storm surge of sea water that comes with major cyclones.

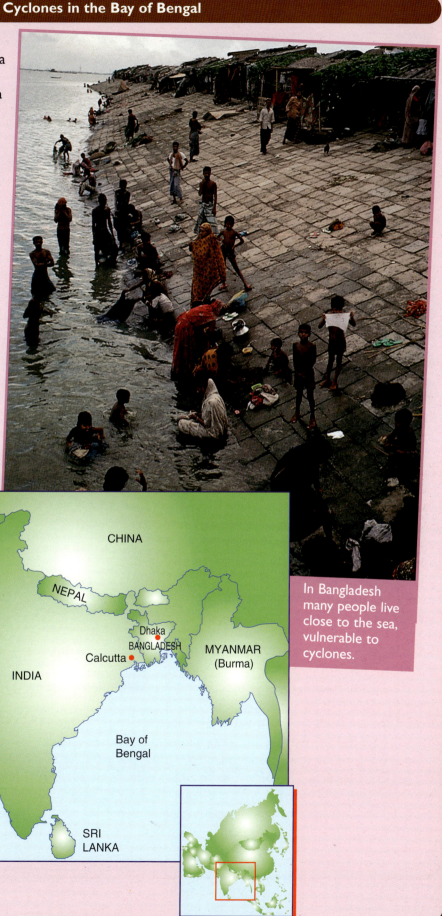

In Bangladesh many people live close to the sea, vulnerable to cyclones.

The track of Hurricane Cesar

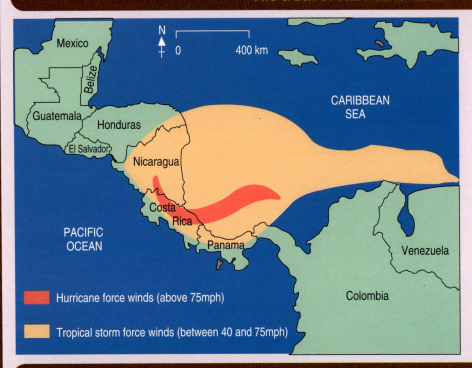

A first hand account of this hurricane is given below. The storm was spotted from satellites operated by the US National Hurricane Centre on the 25th July, 1996. It was first classed as a Tropical Depression and was located between the west coast of Africa and the Caribbean. It quickly increased in strength as it moved westwards and by the 27th July was classed as a hurricane with 70 mph winds. As it crossed Costa Rica and Nicaragua it lost some strength and eventually petered out over the Pacific Ocean off the west coast of Mexico.

A holiday in Costa Rica

Saturday 27th July

Today we followed a trail through the rainforest on the slopes of Arenal Volcano. The volcano coughs into life every twenty minutes, shooting plumes of ash and boulders which bounce down the slopes - thundering and cracking with great power. The clouds burst open just as we reach the edge of the forest. The water pours out of the sky and, three hours later, we emerge drenched to the skin.

   We meet some travellers from the USA. They tell us that a hurricane is heading our way.

Sunday 28th July

We hitch a ride in a jeep along a steep, winding dirt road. Rather worryingly, we encounter several fresh landslides almost blocking the

road. The rain is still torrential. After four hours we reach Monteverde and see that it is almost deserted. The tourists have left and we have the lodge almost to ourselves.

   On the TV, we see that coastal towns have been evacuated for fear of flooding. On the Inter-American highway a bus has overturned and become stuck in a mud slide. Some passengers have escaped but others are trapped inside. In another place the river has risen and burst its banks so rapidly that it has demolished several houses and drowned the occupants before they can escape. In two days they have had the normal rain for two months - even in the rainy season!

Monday 29th July

It is still raining, but not quite as hard. We are travelling to a resort on the Pacific Coast but it is slow going. Many roads and trails are blocked by landslides and fallen trees. Some bridges have been washed away causing long detours. In other places we cross hastily repaired bridges that still seem very shaky and vulnerable.

We hear that 41 people have died in the hurricane and 5,500 homes have been destroyed. The main road south has been badly damaged and will take 2 months to repair. Most power lines in the south of the country have been brought down by falling trees. We are the only people in the resort hotel, everyone else has cancelled or been unable to travel.

In the newspaper, farmers are describing their losses. One reports, 'I had 15 hectares of rice ready for harvest. Now it's all buried under a metre of mud.' For coffee producers, the problem is different. Their crop is safe but the mountain and rural tracks along which they transport the beans are blocked. Without transport, the coffee is worthless.

Talking to the local people, they blame much of the damage on deforestation. Many trees have been cut down to create pasture for cattle. In previous decades, hurricanes, just as severe, have not caused so much damage. But now, with so much of the rainforest cut down, the rain just washes down the hillsides causing landslides and floods. They say that they need flood control schemes and small dams to slow the water before it can cause the damage.

STAGE 3: Drought

In the UK, a dry spell that lasts for 30 days or more is classed as a drought. Other countries use different definitions. Periods of dry weather are quite normal but severe droughts, such as that experienced in the UK through 1995 and 1996, are very infrequent. They can be expected only once in several hundred years.

Some scientists believe that our weather is becoming more variable, and more prone to drought, because of global warming. However, others dispute this view. The issue is considered in more detail in Stage 4.

Droughts, like other forms of severe weather, have a bigger impact on less economically developed countries than on richer countries. We shall see that the UK drought caused hose pipe bans and a rise in the price of some vegetables. Compare this, though, with the famines that killed hundreds of thousands of people following droughts in Ethiopia in the 1980s. However, it should be said that droughts do not cause famine, even in LEDCs, on their own. There is usually some other factor - such as civil war - that prevents food and food aid from reaching the affected people.

In the information that follows, you are given details of droughts in the UK, Spain and southern Africa. Compare the impact of the droughts and make a list of ways in which people overcome the water shortage. You could set it out in the form of a table like the one below:

	Impact of drought	How people overcome drought
UK		
Spain		
Southern Africa		

Record drought relief operation launched

During 1995 and 1996, parts of Britain suffered the longest drought on record. Manchester was drier than Athens! Hose pipe bans were imposed and some businesses (such as car washes) were restricted. The dry weather reduced farm output and the price of some vegetables rose.

Yorkshire Water decided, in November 1995, to run a non-stop shuttle of 1,000 road tankers carrying water from Northumberland to drought stricken areas of West Yorkshire. A year of below normal rainfall had seen 5 reservoirs in Halifax run dry. Stocks of water in reservoirs serving Leeds, Bradford and Huddersfield fell to just 12 per cent of normal.

Although rainfall in Northumberland was also well below expected levels, the region

During the drought many reservoirs fell very low.

has Europe's largest man-made reservoir at Kielder. This giant reservoir holds 44 billion gallons of fresh water. Critics of Yorkshire Water objected to the use of road tankers because of the noise and damage they caused. At the same time, environmentalists opposed further water extraction from the region's rivers, saying that the levels were already too low for fish and plants to survive. They pointed out that leaking underground pipes lost 100 million gallons of water per day. If more investment had been made in mending leaks and creating a water grid to transport water from one region to another, the crisis could have been avoided.

Olive crisis in Spain after worst drought of the century

Between 1990 and 1996, southern Spain experienced a severe drought. Some experts believed that the Sahara Desert was 'moving' north across the Mediterranean.

In Andalucia, olive trees which can normally resist dry conditions, suffered so much that the harvest fell by 50 per cent. Agricultural and processing workers were given food aid to help them survive. The price of olive oil rose rapidly because Spain produces 30 per cent of total world output.

Experts dispute the solutions to the problem. One group suggests that water could be collected and piped from the wet north and north west of the country to the south. The National Hydrological Plan proposed building 270 new dams and hundreds of kilometres of aqueducts. Ecologists object to this solution because of the damage that the new dams, reservoirs and pipes will

Southern Spain experienced a five year drought.

cause in the Pyrenees and north west. The scheme would, of course, cost billions of pesetas of taxpayers money.

A second group of experts suggests that the problem is one of demand, not supply. Farmers have been encouraged to grow fruit and vegetables for the European market using irrigation. 80% of Spain's water is used in this way. Strawberries and rice require much more water than the traditional figs and olives. At the same time, tourist developments have created a huge demand for drinking water, sprinklers on golf courses, fountains and swimming pools.

The result of the drought was for strict rota cuts in supply. At Marbella, on the Costa del Sol, the water was cut for 16 hours per day. Two million people in Andalucia were affected by the rationing. In the long term, the solution is clear. Either spend the money on a national water network or restrict the use of water in the south.

Africans face drought crisis

Terraces and new tree planting in this Tanzanian village help conserve moisture in the ground and also reduce soil erosion when it does rain.

Parts of southern Africa suffered below average rainfall for 14 years between 1982 and 1996. However, because of well organised systems of relief, there was no starvation or famine. This is in contrast to the 1983 and 1984 droughts in the Horn of Africa (Ethiopia, Somalia and Sudan) which triggered mass famine. The affected countries, South Africa, Tanzania, Malawi, Zambia, Zimbabwe, Mozambique, Botswana, Angola, Lesotho, Swaziland and Namibia, tried to coordinate their actions. They predicted, for instance, that the 1995/96 harvest would produce a grain shortfall of 12 million tonnes. They then appealed to the European Union and the United States for this grain and arranged transportation and storage. So, when the time came, there was already a system in place to transport and distribute the food aid. The problem in Ethiopia, a decade earlier, had been that civil war and

inadequate transport made food distribution very difficult.

Farmers in southern Africa are being trained in using techniques to survive the drought. New varieties of grain which grow quicker and withstand drought are being introduced. A water grid, bringing water from wetter, mountainous areas of South Africa and from the Zambezi will allow more irrigation. At the same time, more efficient systems of irrigation, using the 'trickle' method are being encouraged.

Water conservation is extremely important. Ploughing along the contours retains more rainwater than if it is allowed to run off down the slope. Similarly, 'stone lines' laid along the contour of the slope, retain moisture and reduce soil erosion after sudden rain storms.

STAGE 4: Global warming

Most scientists accept that the world's average temperature is rising. We also know that the Antarctic shelf ice sheet is cracking and that parts of the world have experienced exceptional droughts.

It is generally agreed that the rise in temperature has been caused, in part, by increased carbon dioxide (CO_2) emissions. So, in 1992, world leaders met at the Rio Earth Summit (in Rio de Janeiro) and agreed to cut CO_2 emissions, by 2000, to the 1990 level.

Use the information that follows to produce a poster on global warming. You should aim for impact and clarity. On the poster, outline why global warming might become a problem and how people can help by reducing CO_2 emissions.

How will global warming affect us?

For some countries, in particular those forming the Alliance of Small Island States, the issue of global warming is a matter of survival. If rising temperatures cause ice in Antarctica and Greenland to melt, many coastal regions will be flooded. Just consider how many of the world's great cities are at sea level! Island states like Fiji will disappear altogether.

For those of us living well above sea level, what will the effect be? The problem is that no-one knows. Most scientists accept that climate will become more variable, weather conditions will become more severe and farming patterns will be disrupted. The Sahara Desert will 'move' north - but how far? Will the great temperate food growing regions of the world, the American prairies and the North European Plain be affected? Again, no-one knows, but it seems sensible to avoid the risk of a change - if possible.

What causes global warming?

It is generally accepted that global warming is caused by the build up of **greenhouse gases**. The main greenhouse gas is carbon dioxide. It acts like a pane of glass in a greenhouse, letting light energy in but trapping the heat that is created.

Carbon dioxide mostly comes from fossil fuels burnt in power stations and in motor vehicles. It is the developed, industrialised countries of the world that are the biggest producers of CO_2. Coal, oil and gas are vital to our industries and standard of living. Less developed countries claim, rightly, that controls on emissions will prevent them from 'catching up' economically.

Cutting CO_2 emissions is easier said than done. Drastic controls will reduce economic output and could cause standards of living to lower. Therefore, less severe controls are proposed. For example, petrol consumption in cars can be reduced, electric cars or public transport can be encouraged, renewable energy, such as solar, wind and tide, can be subsidised.

Smoke from fossil fuels is blamed for global warming.

STAGE 5: Review

To help you summarise the issues in this Enquiry, you are asked to take the role of a government official in a less developed country. Your task is to draft a plan of how your country could better cope with a particular extreme weather condition (eg, storm, flood or drought).

Outline the problem(s) that 'your' country faces and suggest possible solutions.

Glossary

All the terms listed below are explained in this Enquiry. In each case, write a definition of the term and, if possible, give at least one example.

Global warming

Hurricane

Tropical storm

Drought

Greenhouse gas

Coursework Enquiry

Forecasting the weather

Hypothesis: Weather forecasts are generally accurate.

Many millions of pounds are spent on weather forecasting. Because the UK weather is so variable, it is difficult to predict. Nevertheless, the Meteorological Office claims that it has a very high accuracy for its short term forecasts. This coursework enquiry tests the Met Office claim and helps give an understanding of the factors that affect weather.

It is suggested that you choose a weather forecast that you see or hear every day, such as in a newspaper or on TV. Each day, for at least a month, note the forecast for your area. You will also need to record the actual weather. You might have a weather station at school or, alternatively, you can use the daily weather report in a local newspaper. You can also use personal observation for the duration of precipitation or sunshine.

You need to decide how you will judge the accuracy of the reports. You might, for example, use a three point scale for rainfall forecasts: accurate; accurate but incorrect timing (eg, the rain comes sooner or later than forecast); inaccurate.

Method of enquiry and report writing

1 Explain your hypothesis - why do you think the hypothesis might be true? How has the technology of weather forecasting changed in the last 30 years? (You might include in your reasons: advances in communications, weather satellites and radar.)

2 Decide what data you need - ie, a daily weather forecast for precipitation and temperature (plus wind speed / direction, sunshine?) You also need a daily weather report.

3 Data collection - where will you obtain the data? Will you obtain it purely from secondary sources such as newspapers or TV bulletins or will you carry out first hand observations?

4 Collect the data - will you carry out the coursework on your own or will you share the research with others?

5 Data presentation - you might use tables and graphs to illustrate your findings.

6 Describe and analyse the data - you need to express your findings in written form and say what they show.

7 Accept or reject the hypothesis - you have to judge your hypothesis. What proportion of forecasts are accurate? Are temperature predictions more or less accurate than precipitation predictions?

8 Follow up - use an atlas or other secondary source to obtain the relevant average climatic conditions for your area. Work out the average temperature and precipitation for your month of weather reports. How close are your weather observations to the long term average? Suggest reasons for any variation.
Explain why (and for whom) accurate weather forecasts are important.

7 Natural disasters

to the student

Sudden, natural events can sometimes cause death and destruction. Earthquakes and volcanic eruptions can be particularly dangerous, especially when they occur in densely populated areas. They are so powerful that people cannot prevent them. They are also, in most cases, impossible to predict. However, people living in affected areas are learning ways of minimising the danger. In these Enquiries we look at the causes of earthquakes, volcanic eruptions and avalanches. We also consider where they occur and the ways in which people try to protect themselves.

Tornadoes, lightning, hurricanes and drought can also cause natural disasters. The Enquiry on Severe Weather (pages 110-117) deals with some examples.

questions to consider

1 What causes earthquakes, volcanic eruptions and avalanches?
2 Where do earthquakes, volcanoes and avalanches occur?
3 How can people protect themselves from the dangers?

key ideas

The surface of the Earth is called the *crust*. It is broken into large *plates* which move against each other. This process is called *plate tectonics*. It is the cause of intense geological activity such as earthquakes, volcanoes and mountain building.

An *avalanche* is the sudden downward movement of rock, soil or snow. It can be caused by an earthquake or volcanic eruption or, in the case of a snow avalanche, the cause might simply be the deep build up of snow and a sudden change in weather conditions.

Natural disasters tend to be dangerous because they are sudden and unpredictable. Although huge sums of money are spent on trying to forecast earthquakes, volcanic eruptions and avalanches, little progress has been made so far. Most effort is concentrated on protecting people from the effects of these natural disasters.

activities

Using information from this and the facing page:

a) Name three types of natural disaster.
b) Working with a partner, think back to news events in which natural disasters were reported. Make a list of the ways in which people were affected.

Natural Disasters

EARTHQUAKES

AVALANCHES

Destroy houses
Disrupt communications
Damage crops, factories
 and businesses
Kill and injure people

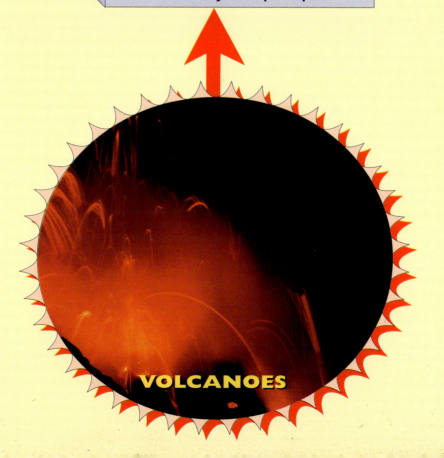

VOLCANOES

enquiry

Earthquakes and volcanoes

In this Enquiry, we consider where earthquakes and volcanoes occur. We shall see that they have a linked cause. By looking at particular examples, we shall see the effects of earthquakes and volcanic eruptions and also see how people try to minimise the dangers.

The outcome of the Enquiry will be an information sheet that you will be asked to produce. It will provide information and advice for a visitor to a region at risk either from an earthquake or volcanic eruption.

STAGE 1: Where do earthquakes and volcanoes occur?

An **earthquake** is a series of shock waves rippling through the Earth's crust. It occurs when rocks move along a crack, or **fault**, after a build up of stress within the Earth's surface. Most earthquakes occur along narrow zones. These zones are at the edges of the plates which form the Earth's crust.

Most volcanoes also occur along these plate boundaries, either where they are moving apart or towards each other. A **volcanic eruption** is when material from inside the Earth is released onto the surface. The material, called **magma**, is solid inside the Earth but becomes a liquid when pressure is released. It can then flow and is called **lava** when it spills onto the Earth's surface.

You can perhaps imagine this as a tube of toothpaste; for most of the time the paste is a solid. When the cap is released and the tube is squeezed, the paste is molten and behaves like a liquid.

The resources which follow contain information on where earthquakes and volcanoes occur. They also explain why they occur.

On a world map, mark and name the location of 5 earthquakes and 5 volcanic eruptions. Draw two sketch diagrams, with labels, that explain why earthquakes and volcanoes occur.

Plate margins with the location of selected major earthquakes and volcanic eruptions

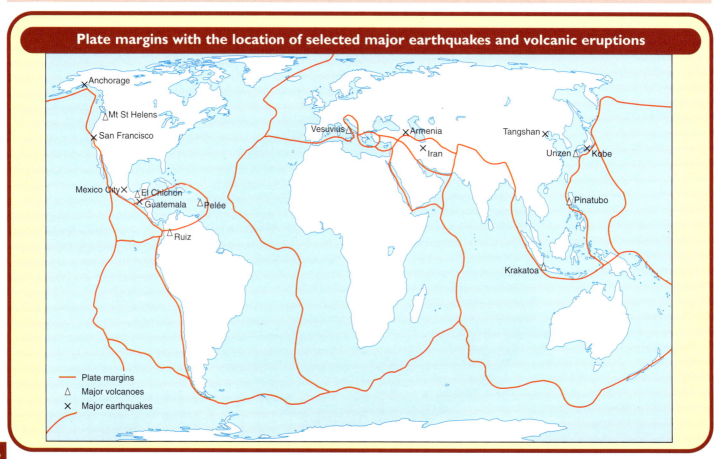

Plate tectonics

This theory explains why volcanoes and earthquakes occur. The Earth's crust is broken into pieces or **plates**. They move - very slowly - in relation to each other because of convection currents within the Earth's mantle.

Where the plates move together, one may sink beneath the other. This is known as a **destructive plate margin**. One example is just to the east of Japan, another lies under the north

The destructive plate margin in the north west USA.

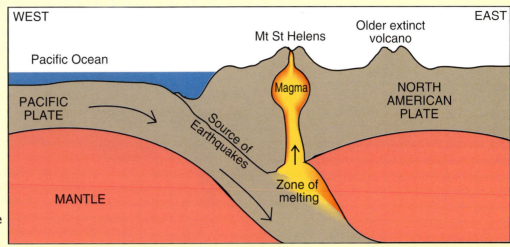

west USA. As the plates move, pressure and stresses build up in the rocks. Then, when a sudden movement occurs there is an earthquake. The point at which the movement occurs is called the focus and the **epicentre** is the point on the Earth's surface above the focus.

Where the plates move apart, lava flows into the 'gap' - so forming a volcanic ridge. This is called a **constructive plate margin**. An example is in the mid-Atlantic where Iceland is formed of lava that has risen up between the plates.

One type of volcano is not found at plate boundaries. This exceptional case occurs where a rising convection current in the mantle causes a **hot spot**. As the overlying plate moves over the hot spot, a chain of volcanoes forms. The best known example is Hawaii. Here is found the world's biggest volcano, Mauna Loa. It rises over 4,000 metres above sea level (and a total of 8,000 metres above the floor of the Pacific Ocean). It is one of the world's most active volcanoes. Its lava runs freely, causing the mountain to have very gentle slopes. Its eruptions tend not to be violent or explosive. Lava can cause danger to people but here it generally runs down channels that are well known - at a speed that allows most people to escape. Mauna Loa is called a **shield volcano** because of its shape.

Diagram to show a constructive plate margin and a hot spot volcano.

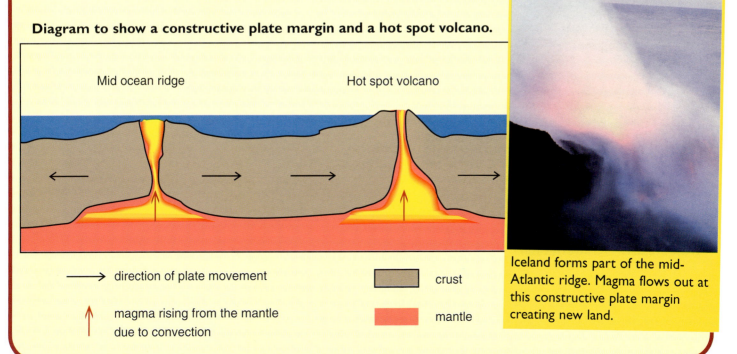

Mid ocean ridge   Hot spot volcano

→ direction of plate movement

↑ magma rising from the mantle due to convection

crust

mantle

Iceland forms part of the mid-Atlantic ridge. Magma flows out at this constructive plate margin creating new land.

Destructive plate boundaries

At a destructive plate boundary, the rocks in the plate which sinks down under the other start to melt. They are less dense than other material in the mantle so rise upwards. When, and if, they reach the surface, a volcano forms. If magma, containing gases, rises under great pressure, or if the flow is blocked and is then suddenly released, the eruption can be explosive. When this happens, large amounts of ash and solid lumps of lava can be thrown great distances into the atmosphere. In the biggest explosions, ash is carried into the upper atmosphere and taken by winds completely around the globe.

When a volcanic eruption is very violent, the whole top of the mountain might be blown away. The resulting crater is called a **caldera**. Our word 'volcano' comes from the island of Vulcano, off the west coast of Italy. This island volcano has a large caldera. It was believed to be the home of Vulcan, the Roman god of fire.

A caldera with a new volcanic cone inside.

Southern California

The San Andreas Fault runs from top to bottom in this photograph.

N

0   300km

PACIFIC OCEAN

San Francisco Bay area

Los Angeles

San Diego

CALIFORNIA

Large eathquakes near the San Andreas Fault

① San Francisco 1906

② San Fernando 1971

③ Coalinga 1983

④ San Francisco/ Loma Prieta 1989

⑤ Northridge 1994

San Andreas Fault ▬

Where plates slide sideways past each other, great pressure and stresses can build in the rocks. An example is in Southern California where the San Andreas Fault forms the boundary between two plates. This fault is 650 miles long and 10 miles deep. The Pacific Plate moves northwards relative to the North American Plate at a rate of 2 inches per year. Along parts of the fault, the plates 'creep' by each other gradually, causing small shocks and tremors. In places where creep is not constant, strain can build up for hundreds of years, producing great earthquakes when it is finally released. In the major earthquake at San Francisco in 1906, the severe shaking lasted for 45 seconds and the San Andreas Fault 'slipped' 15

STAGE 2: What are the effects of earthquakes and how can people minimise the danger?

Although people know where the plate boundaries and fault lines are, it is still not possible to predict exactly when or where the next 'quake will occur. Most of the research is carried out in two places which are at greatest risk - Japan and California. These two places lie along major plate boundaries. They are also extremely rich and developed economies, so have the funds to carry out research and can afford to pay for the building techniques which minimise the risk of damage. However, we shall see, within the last few years, both have suffered earthquakes that have caused death and major damage. In less developed countries, such as Mexico, where people have less money to spend on strengthening buildings, the effects of earthquakes are more severe.

Use the information provided to list the effects of earthquakes. You will be asked in a later Stage to consider how people can reduce the risk of injury and damage

How we measure earthquakes

Earthquakes are recorded by instruments called **seismographs**. They measure the shaking motion of the ground. The **Richter Scale** is used to compare the size, or magnitude, of earthquakes. It is important to note that, for example, Richter Scale 7 is TEN times more powerful than Scale 6, and Scale 6 is ten times more powerful than Scale 5 (and so on).

There are more than 4,000 seismograph stations around the world so information can be recorded and communicated very quickly. Most earthquakes are so small that people do not notice them. In the United States there are 12,000 per year on average. In the UK, a much smaller number occur. We are a long distance from the nearest plate boundary so it is not surprising that UK earthquakes rarely cause any damage.

Although scientists cannot predict when earthquakes will occur, they can calculate their probability. For example, in California, the probability that a quake bigger than 6.5 will occur in the San Francisco Area within the next 30 years is 67%.

The damage caused by an earthquake does not just depend upon its size. The nature of the ground is also very important. On solid rock, the ground shakes much less than on softer ground. Land that is formed from alluvial sands and gravel, for example in an old lake bed, can shake very severely in an earthquake. In the worst cases, the ground becomes like jelly and then most buildings fall down.

Richter Scale	Magnitude	
Great earthquake	≥ 8	Most buildings are destroyed
Major	= 7	
Strong	= 6	
Moderate	= 5	
Light	= 4	
Minor	≤ 3	Hardly felt by humans

The old wooden houses in the foreground survived the 1906 San Francisco earthquake. However, in the downtown area , many wooden buildings did burn down in the fires that broke out.

The large modern buildings in the background are designed to withstand earthquakes.

Official advice for Californians in the event of an earthquake

1 Keep emergency supplies of:

- fire extinguisher
- medications and first aid kit
- portable radio and flashlight with batteries
- bottled water and water purification tablets
- canned and packet food (do not forget pets) and can opener
- camp stove or barbecue for outdoor cooking
- spanners to turn off gas and water supplies.

2 Identify and secure everything in the home that might shake loose or fall. Large items such as fridges and cupboards can move across the floor or topple over. TVs and stereos might fall off shelves.

3 Arrange a meeting point where the family can reunite afterwards. Make sure that everyone knows the emergency procedures.

4 Practice 'ducking', 'covering' and 'staying' under a sturdy piece of furniture. Stay clear of windows, fireplaces, heavy furniture or appliances. Shelter in a hallway or next to an inside wall. Don't attempt to move around the house while it is shaking. Try to wear sturdy shoes to avoid injury from broken glass.

5 Do not turn on the gas afterwards. Do not use matches or lighters until you are sure there is no leaking gas.

6 If you are driving, carefully stop away from bridges, overpasses, trees or power lines. Stay inside the car.

7 If you are already outside, stay outside, get into the open away from buildings and power lines.

8 Do not use the phone except for emergency help. Do not expect the police, firefighters or paramedics to help you quickly. Practice first aid and be aware of hazards.

Collapsing bridges and highways are a major earthquake hazard in California.

Some major earthquakes of the twentieth century

Location	Year	Magnitude
San Francisco	1906	7.7
Anchorage	1964	8.6
Guatemala	1976	7.9
Tangshan	1976	7.8
Mexico City	1985	8.1
Armenia	1988	8.9
San Francisco	1989	7.1
Iran	1990	7.7
Kobe	1995	6.9

You will see from the map on page 122 that these earthquakes occurred on or near major plate boundaries. Although the timing of earthquakes is unpredictable, people living in these areas know that they are at risk.

Modern buildings in California are built with deep foundations and steel frames specially designed to withstand shaking movements.

KOBE QUAKE AFTERMATH: GRIEF AND BITTER COLD

(On January 17th, 1995, a magnitude 6.9 earthquake hit Kobe, Japan.)

Last Tuesday's earthquake turned parts of Kobe, a port city of 1.4 million people, into smouldering ruins without enough electricity, food, fuel or clean water. The death toll is 4,431 with 656 still missing. 10,000 buildings collapsed, over 23,000 are injured and 280,000 are living in emergency shelters. Water is cut off in 1 million homes, 110,000 have no electricity and 900,000 have no gas. With no running water, sanitation is becoming a problem in the emergency shelters.

Many houses were destroyed and people killed by the fires which started when gas pipes fractured. Broken water mains meant that the fire brigade could do little to help.

It was Japan's deadliest earthquake in 70 years. Already the government has accepted that disaster relief plans were inadequate. The scope of the devastation shattered Japanese confidence. They believed that one of the world's most technologically advanced countries could cope with such a natural disaster. Even schoolchildren in Japan are trained to dive beneath the desk at the first sign of shaking.

Kobe losses could reach $50 billion

About 1 per cent of the city's buildings were badly damaged by the quake. This was a far higher percentage than the 1989 quake near to San Francisco. There, a 7.1 scale earthquake had killed 62 people, damaged 18,000 homes and put the San Francisco Bay Bridge out of use for a month.

The death and destruction at Kobe was much higher for a number of reasons. In the old part of the city, many small, closely packed wooden buildings burned once the gas pipes fractured. In San Francisco there are now few wooden houses and they are much less densely spaced.

Kobe also saw the 'pancake' collapse of apartment blocks. In San Francisco these are built with deep foundations designed to cushion shock waves and with steel frames that can withstand shaking. Also, in Kobe, parts of the city are built on alluvial soil and landfill. This causes much more ground shaking than on solid rock; the sandy soil can even flow like liquid.

Although bridges were built with expansion joints so they could shake without collapsing, at Kobe some of the joints failed and some joints were not big enough. American engineers blamed the same faults for the partial collapse of the San Francisco Bay Bridge in 1989, but said that those lessons had now been learned.

In the Kobe earthquake most damage was caused by the fires that broke out.

STAGE 3: What are the effects of volcanoes?

When volcanic lava and ash decompose, they form extremely fertile soil. So around the world, from Java to southern Italy, people are prepared to take the risk and farm in areas at danger from volcanic eruptions. Sometimes they are lucky and the volcano remains **dormant**, ie inactive. Unfortunately, sometimes they are not so lucky. For example, in 1902 on Martinique in the Caribbean, Mt Pelee erupted so suddenly - sending a cloud of heated gas and ash down its slopes - that over 20,000 people died in the town of St Pierre. The Vesuvius eruption in 79 AD sent such thick amounts of ash (5 - 8 metres) onto the nearby town of Pompeii that thousands were suffocated where they slept.

Happily, the biggest explosive eruption of recent times, in 1980 at Mt St Helens in the north west USA, was in a remote area. Sixty one people died. Consider what the effect would have been if a similar explosion took place at Vesuvius where over a million people live nearby in Naples and surrounding towns.

Use the information provided to list the effects of volcanic eruptions. This can be in the form of a line drawing of a volcano with labels attached. You will be asked in a later Stage to consider how people can reduce the risk of injury and damage.

Deadly volcanoes

Volcano	Location	Approximate number of deaths	Year	Major cause of death
Tamboro	Indonesia	90,000	1815	Starvation
Krakatoa	Indonesia	35,000	1883	Tsunami
Ruiz	Colombia	25,000	1985	Lahars / mudflows
Mt Pelee	Martinique	20,000	1902	Gas and ash flow
Unzen	Japan	14,000	1792	Tsunami
Laki	Iceland	9,000	1783	Starvation
Kelut	Indonesia	5,000	1919	Lahars / mudflows
Galunggung	Indonesia	4,000	1882	Lahars / mudflows
Vesuvius	Italy	3,500	1631	Lahars / mudflows
Vesuvius	Italy	3,500	79	Ash flows and falls
Papandayan	Indonesia	3,000	1772	Ash flows
Lamington	Papua New Guinea	3,000	1951	Ash flows
El Chichon	Mexico	2,000	1982	Ash flows
Soufriere	St Vincent	1,500	1902	Ash flows
Oshima	Japan	1,500	1741	Tsunami
Asama	Japan	1,500	1783	Lahars / mudflows
Taal	Philippines	1,500	1911	Ash flows
Mayon	Philippines	1,000	1814	Lahars / mudflows
Agung	Indonesia	1,000	1963	Ash flows
Cotopaxi	Ecuador	1,000	1877	Lahars / mudflows
Pinatubo	Philippines	1,000	1991	Roof collapse/mudflows
Komagatake	Japan	500	1640	Tsunami
Ruiz	Colombia	500	1845	Lahars / mudflows
Hibok-Hibok	Philippines	500	1951	Ash flows

Volcanic hazards

A **tsunami** is sometimes, misleadingly, called a tidal wave. When an underwater volcanic eruption occurs, or an eruption causes a large avalanche into the sea, a massive wave can form. It can travel thousands of miles across an ocean, perhaps less than a metre in height. However, when it reaches shallower water near land, the wave can build in height to 10 or more metres. At Krakatoa, the tsunami was 30 metres high.

Volcanic eruptions have caused mass starvation in the past when crops were destroyed by ash and/or lava. Today this would be unlikely because communications have improved and an international relief effort could be organised.

When large thicknesses of ash fall, the material is loose and unconsolidated. If heavy rains follow, rivers and streams can turn into powerful mudflows. In Indonesia these are called **lahars**.

Sometimes the cause of death is suffocation by ash and gas, often heated, that rushes down a mountainside faster than people can react.

A road blocked by a lava flow.

Lahars or mudflows can cause massive damage.

A house covered by ash.

Krakatoa

One of the largest explosions in historic time occurred in 1883. The island of Krakatoa, lying between Java and Sumatra, virtually disappeared when the volcano erupted. The explosion was heard three thousand miles away and dust was thrown high into the atmosphere where it circled the Earth. As in other major eruptions, the volcanic dust had the effect of reducing the amount of sunshine reaching the Earth's surface. This caused a reduction in average temperatures and was blamed for poor harvests around the world in the following year.

Since 1927 the volcano has been active with a series of small eruptions. A new island, Anak Krakatoa (Child of Krakatoa) has formed.

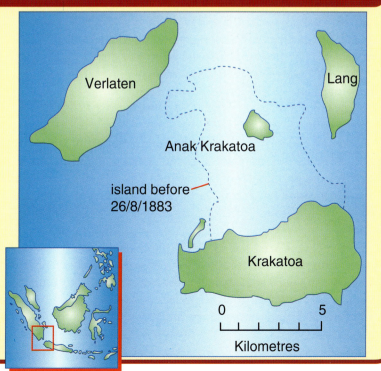

Verlaten

Lang

Anak Krakatoa

island before
26/8/1883

Krakatoa

0   5

Kilometres

Mount St Helens - May 18th, 1980

This volcanic eruption was the most powerful the world had known for 60 years. It was also very well documented because, being in the north west USA, it was accessible to well funded scientists.

After a century of inactivity, the volcano 'awoke' on March 20th with a magnitude 4.0 earthquake. This was followed by two months of intense, small earthquakes during which steam, gases and rocks were ejected. Deep underground, magma (a mixture of semi-molten rock and gas) was rising up into the volcano. This caused a huge bulge on the northern side of the mountain. During April and May the bulge increased by 5 feet per day, until it was 300 feet in size.

On May 18th a magnitude 5.1 earthquake shook the bulge loose, causing the biggest known landslide in historic time. The whole north side of the mountain 'rippled' and 'churned' as it slid down.

The sudden removal of the mountainside led to a release of pressure similar to the removal of a top from a shaken bottle of lemonade. The blast of hot gas, ash and rock destroyed an area of 150 square miles. All vegetation was levelled in the initial blast for a 13 mile distance north of the mountain.

The avalanche of solid material travelled down the slope of the mountain for 15 miles in 10 minutes. It covered a nearby lake to a depth of 180 feet. Part of the avalanche pushed up and over a 1,200 foot ridge and, in places, now covers a 24 square mile area to a depth of 500 feet. Two weeks after the eruption, some of the ash, pumice and other volcanic material in the avalanche flows were still 780° Fahrenheit.

The melted snow from the mountain, together with river and lake water, mixed with the hot volcanic ash to form powerful mudflows or lahars. The largest lahar flowed at a speed of 40 feet per second, destroying everything in its path and leaving deposits 15 feet thick. Downstream, navigation was disrupted on the Columbia River because so much sediment was deposited.

Within 10 minutes of the initial explosion, hot ash had risen 12 miles into the sky. It was blown north eastwards and caused hundreds of small forest fires. For 125 miles the sky appeared dark. Within 2 weeks some ash had blown completely round the Earth. During the nine hours of the eruption, an estimated 540 million tons of ash were deposited over an area more than 22,000 square miles.

Fortunately the area was remote from major towns or cities. Apart from some timber camps and tourist lodges the area was virtually uninhabited. Official warnings and media reports had caused most people to evacuate to a safe distance. Nevertheless, over 60 people were killed, taken unawares by the size of the blast and power of the mudflows.

Mt St Helens.

Trees were flattened by the blast.

The northern side of Mt St Helens was 'blown away' during the eruption.

Montserrat - a Caribbean island

Montserrat lies on a destructive plate boundary. Oceanic crust moving west on the floor of the Atlantic is sinking beneath the plate that forms Central America. The resulting volcanoes create a chain of islands known as an **island arc**. Other examples include Japan and the Aleutian Islands off Alaska.

On July 1996, a puff of smoke told the holidaymakers and residents of Montserrat that the Soufriere Hills volcano was no longer dormant. After hundreds of years of inactivity, steam explosions, ashfalls and earthquakes alarmed the local people. They knew that in 1902 at Martinique, an island further south in the same chain, over 20,000 had died when a glowing cloud of gas and ash rolled down the mountainside.

A team from the United States Geological Survey was asked to help. They wired the mountain with seismographs, tiltmeters and gas analysers to monitor the volcanic activity. When the gas analysers detected sulphur dioxide the vulcanologists knew that it was a sign of magma rising to the surface.

By August 8th the intensity of the tremors caused government officials to evacuate the sick and elderly from the southern half of the island. On August 21st a cloud of ash rose 7,000 feet causing darkness in the island's main town, Plymouth, for half an hour. People became very alarmed but then, after that date, the Soufriere Hills volcano calmed. Since then it has settled to rumbling and hissing steam. Local residents have returned home but the monitoring equipment is still in place. The truth is, no one knows whether the volcanic activity will fade away or a major eruption will occur.

A cloud of ash rose 7,000ft.

STAGE 4: Review

We have seen how earthquakes and volcanoes can cause destruction. People living in more economically developed countries are at least risk of injury because buildings are generally stronger and better built than in poorer countries. At the same time, emergency planning and relief systems are better organised in the richer countries which have the income and resources required.

Your task is to write an information sheet for a UK visitor going to stay for a few months in a major earthquake **or** volcanic area. Use the information from the previous Stages to set out the possible dangers. Then make a list of points of advice for the visitor. Consider what they should be advised to take, the danger signals to watch out for and how to act in an emergency.

Glossary

All the terms listed below are explained in this Enquiry. In each case, write a definition of the term and, if possible, give at least one example.

Crust	**Magma**	**Shield volcano**
Plate	**Lava**	**Seismograph**
Plate tectonics	**Constructive plate margin**	**Richter Scale**
Avalanche	**Destructive plate margin**	**Dormant volcano**
Earthquake	**Epicentre**	**Tsunami**
Fault	**Caldera**	**Lahar**
Volcanic eruption	**Hot spot**	**Island Arc**

Enquiry

Avalanches

We have seen in the previous Enquiry that earthquakes and volcanoes can trigger avalanches of rock and mud. In this Enquiry we look at snow avalanches. They occur mainly in winter in mountainous regions. We shall consider why they occur and what effect they have on human activity.

The outcome of the Enquiry will be a guide that you will be asked to write for winter mountain users.

STAGE 1: When and where do snow avalanches occur?

It is obvious that for a snow avalanche to occur, there must first be snow. Clearly then, mountain areas in winter are the place that avalanches will be found. Mountains have colder weather than low lying areas (see page 99 for an explanation) and have the steep slopes required for avalanches.

Although all mountain regions experience avalanches, they affect most people in those areas which have big winter populations. These include the ski resorts of the Alps, Pyrenees, Norway, the USA and Canada.

Mark on an outline map of Europe the location of the Alps and Pyrenees. Make notes on the map that state when most avalanches occur and why they affect people at that time of year.

Avalanche frequency in the French Alps

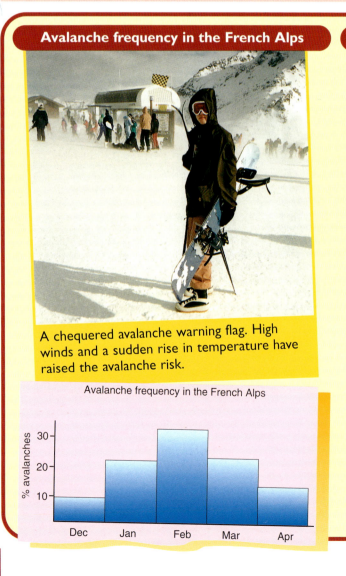

A chequered avalanche warning flag. High winds and a sudden rise in temperature have raised the avalanche risk.

Avalanche frequency in the French Alps

[Bar chart: % avalanches on y-axis (10, 20, 30), months Dec, Jan, Feb, Mar, Apr on x-axis. Dec ~9, Jan ~22, Feb ~32, Mar ~22, Apr ~13]

The Alps and Pyrenees

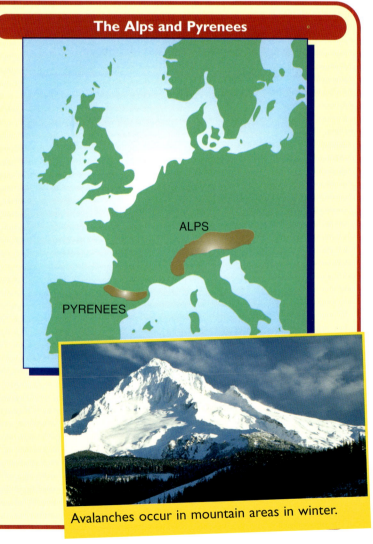

ALPS

PYRENEES

Avalanches occur in mountain areas in winter.

STAGE 2: Why do avalanches occur?

We have seen that the occurrence of snow avalanches requires snow and slopes. The precise causes of an avalanche can be divided into 'background reasons' and 'trigger events'. The information which follows describes these conditions.

Use the information to draw a spider or star diagram that explains why avalanches occur.

Background conditions

Avalanches occur when snow on a slope is not strong enough to remain stable. This requires a build up of snow. Often snow forms layers after each fall, each layer compressing the layer underneath. The deeper and more prolonged the snowfall, the more unstable will be the snow.

> The angle of slope is important:
>   < 20 degrees, the snow remains stable
>   > 60 degrees, the slope is too steep for snow to accumulate.
> So, slopes between 20 and 60 degrees are most prone to avalanches.

Forest cover is the third crucial factor. Trees can help trap snow and slow down the speed of falling snow. They therefore reduce the chances of avalanches starting.

Trigger events

- A sudden rise in temperature can cause snow to become unstable. Melting water can trickle between the layers of old snow and loosen the top layer.
- Heavy fresh snowfall on layers of old snow can become unstable simply due to gravity.
- Wind can cause the top layer of snow to form 'slabs' which are unstable and liable to slip down.
- Vibration from a skier, a falling rock or road traffic can all set snow moving.

An avalanche.

Dense tree cover prevents avalanches.

Types of avalanche

Avalanches

Powder (loose) avalanches
- dry snow
- accompanied by strong blast of air as snow falls
- usually quite small
- low density snow
- rapid speed 20 -70 m/sec

Slab avalanches
- snow moves as a block
- high density snow
- small to very large
- moderate speed 5 - 30 m/sec
- wet snow

STAGE 3: How do people cope with avalanche hazards?

Every year, avalanches kill skiers and mountaineers. They also damage and destroy buildings, farms, roads and power lines.

In ski resorts, where thousands of people depend upon tourism for a living, the danger from avalanches is also an economic risk. If people die or are injured it is bad for business, so every care is taken to minimise the danger.

The resources which follow outline some of the methods used to minimise the risks. Make a simple sketch drawing of an Alpine scene. Add labels to show the methods used to reduce avalanche hazards.

'Second skier dies in Alps'

An avalanche in the Alps killed a second Briton in two days. The skier was off-piste (ie, off the prepared ski runs) and was not using an automatic avalanche alarm. These emit electronic signals and help rescuers locate buried people quickly.

Throughout the Alps a four-point avalanche warning was in force. The highest category of alert is a five-point warning. Mild weather combined with strong winds made snow conditions hazardous. The Ski Club of Great Britain warned skiers not to go off-piste while weather conditions made snow fragile and dangerous. A spokeswoman said: 'If the wind compacts the snow, you can get a 'slab' effect where vast slabs of snow are moved by the wind and cause an avalanche'. She added that the particular ski area was notorious for avalanches because so much of the natural forest had been cleared from the slopes to make wider ski runs.

- *newspaper report, January 1996.*

Avalanches are likely when there are gaps in the tree cover. Where trees have been cleared to make ski runs there is extra danger.

Steel fences and barriers reduce the risk of avalanches.

Avalanche warnings and precautions

All ski resorts operate a warning system if avalanches are threatened. They spend a great deal of money on weather forecasting and monitoring of snow conditions. When the risk is most severe, lifts are closed so that people cannot get up the mountains. However, conditions can change suddenly and, for some people, skiing and climbing are adventure activities which are worth the risk. Sensible people do not venture out alone and they tell others where they are going.

Avalanche danger. Do not ski off the prepared piste.

After heavy snow, many resorts actually set off small, controlled avalanches to prevent a bigger build up of snow. Detonators and mortars are used to remove snow that is overhanging ski pistes, roads and rail lines. Snow fences and barriers are often built on the top of slopes to prevent the build up and movement that could start an avalanche. Trees are also planted on bare slopes for the same reason.

The most vulnerable roads and rail lines are enclosed in concrete boxes which allow avalanches to pass over them harmlessly. Triangular mounds and concrete posts are built uphill of vulnerable structures such as pylons and isolated houses. Local villagers know from experience where most avalanches occur. They are therefore careful to build their houses, villages and ski resorts in the safest locations.

STAGE 4: Review

Using the information you have gathered, produce a flow chart that illustrates the factors that caused the avalanche described in the newspaper article: 'Second skier dies in the Alps'. Set out the chart in a series of boxes that describe the particular long and short term factors in this case. In other words, show how the 'background conditions' and 'trigger events' combined to cause the avalanche.

Glossary

All the terms listed below are explained in this Enquiry. In each case, write a definition of the term and, if possible, give at least one example.

Powder avalanche   **Slab avalanche**

8 Managing physical systems

to the student

In this Enquiry, we shall consider two physical systems - rivers and coasts. We shall look at the processes that operate within these systems and the landforms they create. Physical systems affect people's lives. For instance, floods can damage crops and property. Severe floods can drown people and cause disease if drinking water becomes contaminated with sewage. There are also positive features of physical systems. Rivers provide water for irrigation and hydro-electric power. Coasts provide scenic landscapes that are used for leisure and recreation. For all these, and many other reasons, the management of physical systems is important. We shall look at particular examples of rivers and coastlines to see how people try to control them.

questions to consider

1 What is a physical system?
2 How do river and coastal systems work?
3 What landforms are created by the processes at work in rivers and on coasts?
4 Why do people try to manage physical systems?
5 What methods do people use to manage river and coastal systems?

key ideas

A *physical system* involves *inputs, processes and outputs*. For example, in a river system, rainwater is an input and the river flowing into the sea is an output. Processes within the system, such as erosion, create landforms like valleys and waterfalls.

People try to manage physical systems because they are useful - for example, rivers provide drinking water and energy; coasts provide leisure activities. However, physical systems can sometimes be dangerous - floods are the best example - and this is another reason for managing them.

Because rivers and coasts can be used in many different ways, conflict sometimes arises over how they are managed. This is often true of multipurpose projects. These are development schemes which bring a number of benefits to an area.

activities

Using information from this and the facing page:

a) Draw a simple flow diagram that shows, with examples, the elements in a physical system (ie, input, output and processes).

b) Working with a partner, make a list of all the ways you can think of that people use rivers and coasts.

PHYSICAL SYSTEMS

Waves

Precipitation

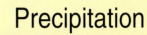

INPUTS

Erosion and deposition

Erosion and transport

PROCESSES

Coastline

OUTPUTS

River flows into the sea

enquiry

Why manage rivers and coasts?

Rivers and coasts are two examples of physical systems. There are many others but these two have a major effect on our lives. In this Enquiry, we shall look at the processes at work within the systems and the landforms they produce. However, the focus of the Enquiry will be the study of how rivers and coasts affect people and how, in turn, people try to manage them.

The outcome of the Enquiry will be a decision making exercise on managing one particular stretch of coastline within the UK.

STAGE 1: How do river systems work?

The water that flows through rivers is one part of a greater cycle that is occurring all around us every day - the **Hydrological or Water Cycle**. Water evaporates from the sea, lakes and rivers, rises into the atmosphere and forms clouds. Different forms of precipitation such as rain, sleet and snow fall from the clouds to the ground. Some of the rain flows along the ground as **surface run-off**. This finds its way into rivers and the cycle is complete. Some rain seeps into the ground, in a process called **infiltration**. A proportion of this **groundwater** finds its way back into the water cycle and some is used as drinking water when people sink bore holes and wells.

A river starts with precipitation and surface run off. It ends as it enters the sea or a lake. In between, the river creates landforms as it erodes, transports and deposits material downstream. The input into a river system is precipitation. The output is the water flowing into the sea.

The resources that follow show the processes at work within a river and the effects they have. Draw a labelled diagram of the hydrological cycle. List the inputs, processes and outputs of the cycle in a table (like that shown below), together with a brief explanation of the terms.

Hydrological Cycle		
INPUTS	**PROCESSES**	**OUTPUTS**
Precipitation - falls as rain or snow onto the ground and vegetation and into rivers.	Infiltration - water moves down through the ground.	River run-off - water flows into the sea.

Make a list of processes at work in rivers and, in each case, briefly write what each one does. Similarly, make a list of the landforms that are created by river processes and describe how each is formed.

Finally, draw a sketch diagram of a drainage basin and label the watershed, the source, a tributary and the catchment area. Add a note that explains what a storm hydrograph shows.

Rivers form one part of the hydrological cycle.

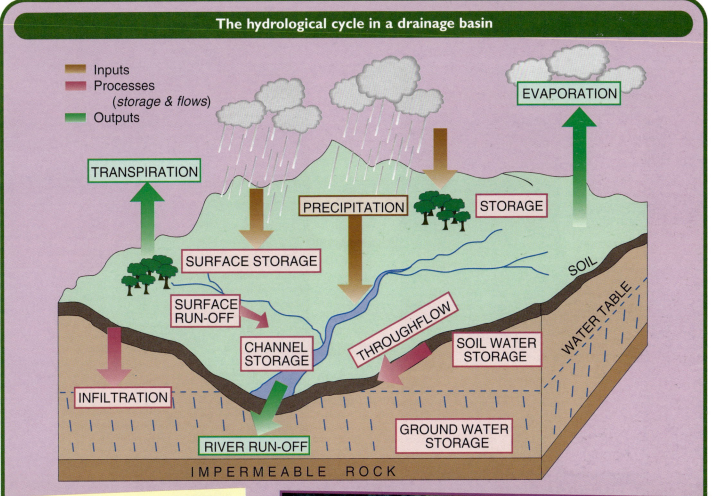

The hydrological cycle in a drainage basin

Inputs
Processes
(*storage & flows*)
Outputs

TRANSPIRATION
EVAPORATION
PRECIPITATION
STORAGE
SURFACE STORAGE
SURFACE RUN-OFF
CHANNEL STORAGE
THROUGHFLOW
SOIL WATER STORAGE
SOIL
WATER TABLE
INFILTRATION
RIVER RUN-OFF
GROUND WATER STORAGE

IMPERMEABLE ROCK

Terms used in the hydrological cycle

Drainage basin - the area of land which drains into a particular river.

Transpiration (or Evapo-transpiration) - transfer of water from plants to the air.

Evaporation - transfer of water from the ground to the air (as it changes from a liquid to a gas).

Infiltration - movement of water into soil and rocks.

Throughflow - water flowing through the ground into rivers and streams.

Groundwater - the name given to throughflow water.

Surface run-off - water running along the ground into streams.

Water table - the level below which rocks are saturated with (ie, full of) water.

Storage - water held within the system.

Trees play an important role in the hydrological cycle. Their roots, stems and leaves store water. Some of this is lost directly back into the atmosphere in the process called transpiration. The root systems slow down surface run-off and cause more water to seep down into the soil and rocks. When trees are cleared (ie, deforestation occurs), surface run-off is quicker. This can cause soil erosion, particularly on steep slopes and in regions where rainstorms are heavy and sudden.

EROSION

Hydraulic Action
Force of water undercuts and removes material

Corrasion
Rocks and boulders exert pressure on sides and bottom of channel - loosening material

Attrition
Rocks collide and break up into smaller pieces

TRANSPORT

Solution
Dissolved material (invisible) carried along

Suspension
Small particles can be carried along in the flow

Saltation
Bedload particles are bounced along the river bed

Traction
Particles are rolled along the river bed

DEPOSITION

Deposition
River slows and dumps transported material to form sand banks and mudflats. Alluvium is also deposited on the valley floor after floods

SEA

River cliff
The river is eroding and undercutting the outside bend of the meander. This is because the water is flowing fastest and eroding most strongly at this point. On the inside of the bend the water is flowing less fast and deposition occurs. The gentle slope on the inside of a meander is called a slip-off slope.

Boulders in a stream
Pebbles, rocks and large boulders are transported when a river flows strongly.

The mouth of a river
Silt and sediment are deposited when the river slows down as it reaches the sea.

Waterfalls

Waterfalls are often found in steep sided valleys. They are formed when a river erodes down to a layer of rock that is particularly resistant to erosion. Rocks below the lip of the waterfall are eroded in an undercutting process. As the waterfall wears away the rocks, it 'retreats' - leaving a gorge.

The deep water at the foot of the waterfall is called a 'plunge pool'.

Bedload

This dry river bed clearly shows the bedload that is transported in normal times when the river is flowing. Only at times of peak flow, for example in times of flooding, will the largest rocks and boulders be moved downstream.

Ganges delta in Bangladesh

When a river that carries a large amount of silt is slowed down (as it reaches a lake or the sea), it deposits its load. A delta of braided channels then forms. The channels are called distributaries.

Sometimes the river forms natural embankments. They are called levees in the Mississippi Valley of the USA. The embankments are formed when silt is deposited on the river banks at times of flood. The river level can then sometimes be above the surrounding land. Often, the embankments are strengthened to protect the land from flooding.

EROSION

TRANSPORT

DEPOSITION

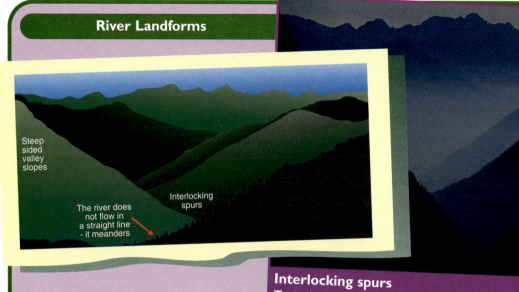

Steep sided valley slopes

Interlocking spurs

The river does not flow in a straight line - it meanders

Interlocking spurs

Typical of a river near its source. Rivers very rarely flow in a straight line, instead they meander in a series of bends. As the river erodes downwards it causes the valley sides to appear as interlocking slopes or spurs.

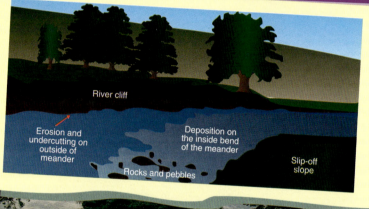

River cliff

Erosion and undercutting on outside of meander

Deposition on the inside bend of the meander

Slip-off slope

Rocks and pebbles

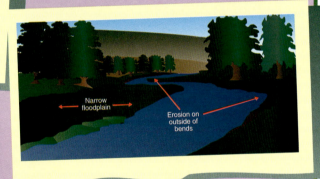

Narrow floodplain

Erosion on outside of bends

Meanders

Typical of a river in its middle reaches. The process of erosion on the outside of the bends and deposition on the inside of the bends causes the meanders to move down the valley. A small flood plain has formed as the river meanders have moved.

If the river starts to erode more strongly it will cut down to a new level and leave the flood plain as a **terrace**. This process can occur several times - leaving a succession of terraces.

The river has deposited sediment on its bed - so raising it above the surrounding land

Flood plain lower than river level

Artificial embankment built to prevent flooding

Flood plain and embankments

Typical of a river in its lower reaches or near its mouth. These embankments have been built to protect the surrounding land from flooding. The river has deposited silt on its bed and, in this case, is flowing at a level above the fields on each side. Although in the short term the farmland is safe from flooding, if a major flood occurs the water will cover a wide area and will not drain back into the river.

The river is brown coloured because it is transporting so much silt.

Cut-off or ox-bow lake

Erosion on the outside of these meanders will cut through the neck of the meander - so forming a new cut-off

Ox-bow lakes: Kuskowin River, USA

Typical of a river in its lower reaches. The river erodes its banks on the outside of meanders because that is where the current is flowing fastest. It sometimes erodes sideways (or laterally) so much that it cuts through the narrow neck of land, leaving a 'cut - off' or ox-bow lake. The river is a browner colour than the lakes because of the large amount of silt that it is transporting.

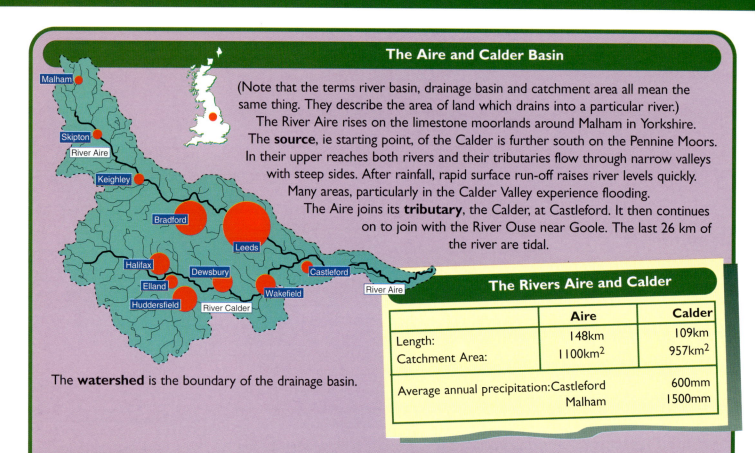

The Aire and Calder Basin

(Note that the terms river basin, drainage basin and catchment area all mean the same thing. They describe the area of land which drains into a particular river.)
The River Aire rises on the limestone moorlands around Malham in Yorkshire. The **source**, ie starting point, of the Calder is further south on the Pennine Moors. In their upper reaches both rivers and their tributaries flow through narrow valleys with steep sides. After rainfall, rapid surface run-off raises river levels quickly. Many areas, particularly in the Calder Valley experience flooding.
The Aire joins its **tributary**, the Calder, at Castleford. It then continues on to join with the River Ouse near Goole. The last 26 km of the river are tidal.

The **watershed** is the boundary of the drainage basin.

The Rivers Aire and Calder

	Aire	Calder
Length:	148km	109km
Catchment Area:	1100km^2	957km^2
Average annual precipitation: Castleford		600mm
Malham		1500mm

Storm hydrograph and rainfall graph - River Calder, August 1996

The hydrograph shows the flow of water in the Calder at Elland. The flow is measured in cumecs - cubic metres per second. Just after 17.00 hrs, the flow of the river increased dramatically. This was due to a number of factors.

The rainfall graph shows that just before the river rose there was heavy rainfall on the moors above Elland. This fell suddenly in a short thunderstorm. The region was in the middle of a long dry period and the ground was very hard. The combination of heavy rain and hard ground meant that most of the water flowed as surface run-off. The steep sided, narrow valleys of the upper Calder also increased the rate of run-off.

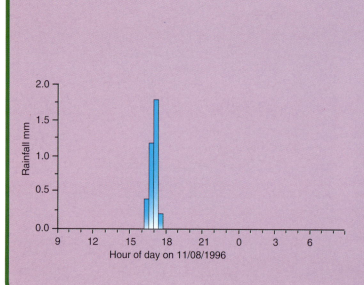

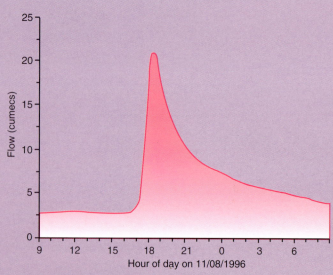

STAGE 2: The River Rhone

The Rhone flows through Switzerland and France to the Mediterranean. The Rhone's drainage basin contains some of the most prosperous and economically developed regions of Western Europe. People use the river and its tributaries for transport (navigation by barges), hydro-electric power, drinking water, recreation, irrigation and water for manufacturing industry. Along the banks of the river are major cities such as Geneva and Lyon. It is therefore very important that the river is managed in a way to serve all the demands made on it. At the same time, control measures have been put in place to prevent flooding which could disrupt such a prosperous region.

The resources which follow contain information on the Rhone basin and the Multipurpose Project which has been developed to manage the river. This type of scheme aims to develop the usefulness of the river and its valley in several ways.

Write a brief note that describes the problems faced by people living in the Rhone Valley before the multipurpose scheme was put in place.

On a sketch diagram make notes on how the river and valley are now used (after the scheme) and how the river is being managed.

'Le Rhone Sauvage'

'Le Rhone Sauvage' literally means the 'wild Rhone'. The name was given to the river because of its fast flowing currents and dangerous floods. The river has always been an important cargo route to and from the Mediterranean. But navigation was difficult because of the river's meanders, its sand and gravel banks and the changing water levels throughout the year. Because the river is fed by the glaciers and snowfields of the Alps, at the time of the spring thaw the water level rises rapidly. During winter, when the Alps are frozen, water levels are much lower.

The Rhone is fed by meltwater from the Alps.

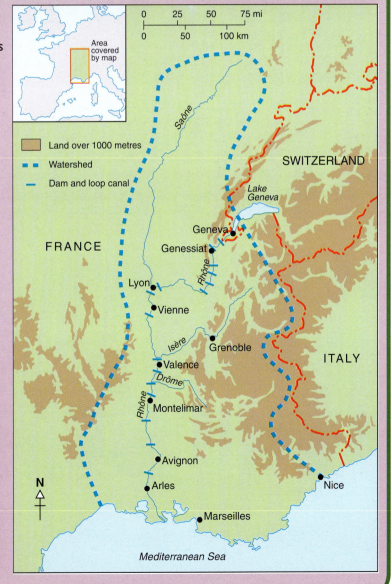

In 1933, the National Rhone Company (Compagnie Nationale du Rhone) was set up to control the river by building dams. The largest was built at Genessiat. It was completed in the 1940s and there are now 12 other major dams along the French section of the river. These dams and smaller barrages not only control the flow of the river but provide water for hydro-electric power stations (HEP). Over 20 per cent of France's electricity comes from the Rhone. Locks and loop canals allow shipping to bypass the dams.

Dams have been built to regulate the flow of water. At times of flood the water is held back. During dry spells, water is released. Hydro-electric power is generated using water from the dams.

The availability of safe, cheap water transport, together with the HEP, has led to increased industrialisation.

River ports such as Valence and Lyon have developed industries including chemicals, engineering, cement making, flour milling and oil refining. Barges provide much cheaper transport for bulky raw materials than either road or rail.

The reduced danger of flooding on the flood plain has increased the area of useful farmland. Irrigation, using water from the river, has increased crop yields and allowed commercial rice growing on the Rhone delta. Much of the flood plain is now used for intensive commercial agriculture growing fruit, vegetables and tomatoes. Previously, only olives and winter wheat could be grown without irrigation. Some of the natural wetlands of the Camargue, on the delta, have been preserved as nature reserves.

At Lyon, Port Edouard Herriot has been built to handle the river traffic. Industries such as oil refining and chemicals have grown up near the dock sides.

A view across the Rhone showing Lyon and the river side wharves.

Sketch cross-section of the Rhone Valley showing the problems before the river was controlled.

WEST

EAST

Massif
Central

Flash floods after
summer storms

Village sometimes
floods

Alps

Snow melt in
spring causes
floods

Fertile flood plain but
often floods in spring,
dry in summer

Braided channel in summer
Less than 1 metre in depth

The Camargue is an area of wetland in the Rhone delta. Some parts have been developed for commercial rice cultivation but large areas have been protected as natural wildlife refuges.

Irrigation in the Rhone Valley. A large variety of crops, including fruit and vegetables, are grown on the fertile valley soils. Previously, before the river was controlled, the late summer drought and spring flooding made farming difficult.

A coastline contains the same elements of a system that we have seen in rivers. The inputs are waves and the landscape before erosion. Waves are formed when wind energy is transferred to seawater or, in other words, when the wind blows over the sea. The processes of erosion, transport and deposition are the same as those in rivers. The output from the system are the coastal landscape features such as beaches and cliffs.

The resources which follow describe the processes and landforms found along coastlines. Make a list of the processes and landforms and write a brief note on each.

Then describe how people try to manage the processes of erosion and deposition along the coastline. Explain why they do this.

Coastal landforms caused by erosion

Chalk is resistant to erosion so forms high cliffs

Stack

A free standing column of rock left standing as the cliffs have eroded

Chalk cliffs

The sea erodes and undercuts the cliffs

Chalk cliffs and stack - Flamborough Head

Chalk is resistant to erosion so, although the sea is eroding the base of the cliffs, they do not quickly collapse. However, over time, the cliffs do 'retreat'. Here a small section of cliff has been left standing as a stack. It will eventually be eroded at its base and collapse.

The sea erodes the base of the cliffs by hydraulic action and by corrasion. Hydraulic action is simply the force of the water as waves break against the rock face. Tremendous pressure is exerted as the water is forced into cracks and joints. Corrasion occurs when pebbles and rocks are hurled by the waves at the foot of the cliffs.

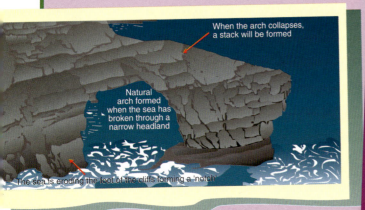

When the arch collapses, a stack will be formed

Natural arch formed when the sea has broken through a narrow headland

The sea is eroding the foot of the cliffs forming a 'notch'

Natural arch - Pembrokeshire

The arch is formed as the sea erodes both sides of a narrow headland. Eventually the sea breaks through - often by eroding and widening a point of weakness such as a fault or joint in the rock. Over time, as the arch is widened by erosion, it will eventually collapse leaving a stack.

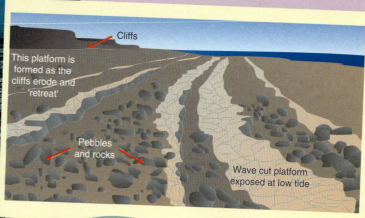

Cliffs

This platform is formed as the cliffs erode and 'retreat'

Pebbles and rocks

Wave cut platform exposed at low tide

Wave cut platform

When the sea erodes tough and resistant rocks, cliffs form. Over time these retreat, leaving behind a wave cut platform. This is normally only exposed at low tide. It consists of rock pools and bands of resistant rock. It eventually erodes as waves and tides wash pebbles and rocks backwards and forwards across it.

The area as it is today - from above.

The area as it is today - from below.

Coastal erosion at Scarborough

In 1993 a combination of storms undercutting the sea cliff and heavy rains saturating overlying clay caused a sudden landslide on Scarborough's South Cliff. An estimated 1 million tonnes of material slipped downwards - extending the foreshore by 100 metres. The Hollbeck Hall Hotel, standing on the top of the cliffs, was so badly damaged that it had to be demolished.

The site has now been landscaped and trees have been planted to stabilise the slope. Over 30,000 tonnes of large boulders have been imported to act as a breakwater at the foot of the cliffs - at a cost of £2 million. This expensive solution to coastal erosion is only worthwhile in built up areas where there is a lot of expensive property. Further south along the Yorkshire coast it has been decided that it is not worth the money to try and save farmland, isolated farmhouses and holiday bungalows.

Beaches and Longshore Drift

Wave action can transport material along a beach in a process known as longshore drift. Depending on the particular circumstances, this can remove sand and pebbles from one area and redeposit them elsewhere. Waves can also sort and grade material on a beach. When a wave breaks on a beach the upward movement is called the 'wash'. The water running back down the beach to the sea is called the 'backwash'. The wash is stronger than the backwash and it transports larger pebbles and rocks up the beach. The weaker backwash transports smaller and lighter particles down towards the sea. For this reason, on sandy or pebble beaches, the larger rocks and particles are found higher up the beach than the smaller particles.

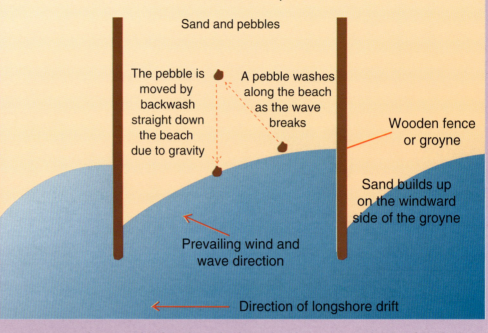

Under gentle conditions, constructive waves wash material up the beach. In storms, destructive waves remove sand and pebbles.

Sand and pebbles

The pebble is moved by backwash straight down the beach due to gravity

A pebble washes along the beach as the wave breaks

Wooden fence or groyne

Sand builds up on the windward side of the groyne

Prevailing wind and wave direction

Direction of longshore drift

Clay cliffs - soft and easily eroded by the sea

Groyne

Protective wooden sea 'wall' to slow erosion of the cliffs

Groynes along the Norfolk coast

Groynes are wooden fences designed to slow the process of longshore drift. They are usually built by councils who wish to protect the shoreline from erosion. This is important for holiday resorts which do not want to lose their beaches.

Sea

Loe Bar formed across the inlet

The Loe

Loe Bar - Cornwall

On the south coast of Cornwall and Devon there are a number of deep inlets called rias. These are drowned river valleys - formed when the sea level rose. One of these inlets, at Loe, has been completely blocked by sand and pebbles. This has been transported along the coast by longshore drift. When an inlet or bay is blocked in this way the feature is known as a **bar**.

N

Porthleven Sands

The Loe

Loe Bar
Longshore drift has transported sand and pebbles across the mouth of the inlet to form a bar

English Channel

0   1 km

This part of the Yorkshire coastline is retreating. Material is being eroded and transported southwards.

Direction of longshore drift

River Humber

NORTH SEA

Spurn Head

Spit formed where the Humber Estuary breaks the coastline

N

0   50 km

Spurn Head

A combination of soft clay cliffs, strong southerly currents and North Sea storms has caused the Yorkshire coastline to retreat several miles since Roman times. There are a number of 'vanished' villages far out to sea.

The eroded material is transported south by longshore drift to the mouth of the Humber where it is deposited to form Spurn Head. The flow of water from the Humber prevents the sand and pebbles from completely blocking the estuary. Spurn Head is an example of a **spit**.

Tombolo

When wave action from two different directions causes a sand bar to form across from a mainland to an island, the feature is known as a **tombolo**.

Druridge Bay is a seven mile stretch of Northumberland coastline bordered by wide sandy beaches and sand dunes. Over the past 35 years, management of the bay has involved a variety of uses. Human users have created many pressures on the area causing changes to the natural environment.

The resources which follow give information about the location and landscape features of the bay area. Also included are the different groups of people and organisations which have been involved in using and managing this area of coastline.

Make a list of the users (past and present) of the Druridge Bay area. Note any conflicts between the interests of the different users (for example, tourism versus open cast mining).

The Druridge coastline - recreational land use

Druridge Bay is a 7 mile stretch of sandy beach and sand dunes. It is a coast of deposition. In other words, the processes of deposition are stronger here than the processes of erosion. The result is a wide beach and, because the sand has blown inland, there are extensive sand dunes.

Druridge Bay Country Park, which opened in 1989, was a joint venture between British Coal and the Countryside Commission. The area was formerly the site of open cast coal mining. Now the main extraction area has been flooded and forms Ladyburn Lake, used by windsurfers and other watersport enthusiasts.

The Park and adjoining beaches are popular areas for leisure and tourism all year round, but especially in July and August. The large cities of Newcastle and Sunderland lie less than 30 km to the south so many day visitors are able to enjoy the beautiful scenery and quiet atmosphere. In the Park, a Visitor Centre, car parks, picnic sites, footpaths and bridleways have all been provided.

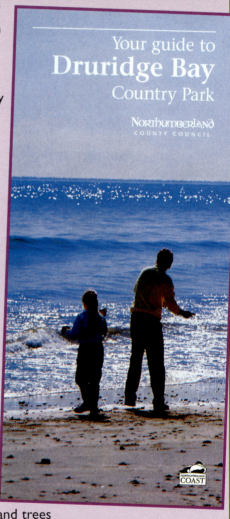

Your guide to
Druridge Bay
Country Park
NORTHUMBERLAND
COUNTY COUNCIL

Plantations of trees have been established in the Park to improve the visual attractiveness and also to encourage wildlife. The lake and beach are used by large numbers of ducks and other migratory birds such as dunlin and turnstone. The dunes are a Site of Special Scientific Interest (SSSI). 283 species of flowering plants have been recorded as well as butterflies and other insects.

The environmental importance of the Bay has been recognised by national and international agencies. In total there are five SSSIs and three Wildlife Trust nature reserves as well as the Country Park. The Bay has been designated a Heritage Coastline, a European Special Protection Area and a Wetland of international ranking.

The Bay is used for educational purposes. Organised walks and field excursions encourage young people to become interested in nature conservation and environmental issues.

Inland, both arable and pastoral farming takes place. Wind breaks of fencing and trees are used to protect the land from wind blown sand. These shelter belts also reduce crop damage from the occasionally strong north east gales.

One problem that has arisen with greater recreational use is the issue of sewage treatment. In the middle of the Bay there is a sewage pumping station that serves the surrounding area. Sewage is screened for solids but otherwise is not treated. Most of the waste flows directly into the sea but some has also contaminated Ladyburn Lake - posing a health risk to people there.

Druridge Bay

Druridge Bay - industrial land use

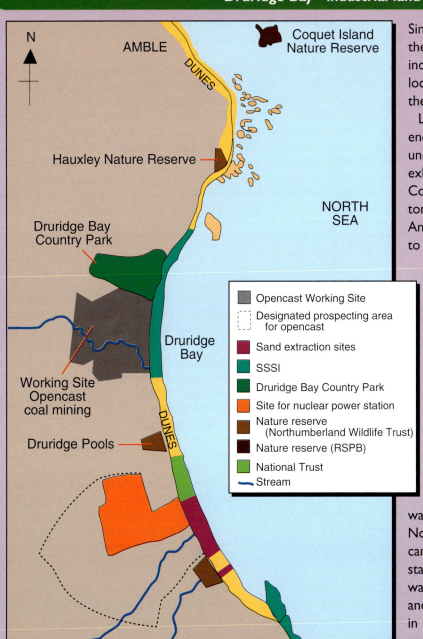

Legend:
- Opencast Working Site
- Designated prospecting area for opencast
- Sand extraction sites
- SSSI
- Druridge Bay Country Park
- Site for nuclear power station
- Nature reserve (Northumberland Wildlife Trust)
- Nature reserve (RSPB)
- National Trust
- Stream

0   0.5 km

Since 1959, sand has been extracted from the beach for use in the construction industry. After a long running campaign by local people, backed by the National Trust, the extraction was ended in 1996.

Large reserves of coal, at depths shallow enough to allow open cast working, exist under the Bay. One large site has been exhausted and restored - to form the Country Park. A total of nearly 3 million tonnes was extracted from this one site. Another open cast mine still operates just to the south of the Park and an area further south has been designated as a possible future site. Many local people, and the regional economy, depend upon the supply of the relatively cheap, high grade coal that is extracted in the area.

In 1978 it was announced that a nuclear power station might be built at Druridge Bay, partly because the area is relatively unpopulated and remote. The site is ideal because it provides a large area of flat land for building, it is relatively remote from large towns and cities - so easing safety fears, and the coastal location allows discharge of hot water (from the turbines) into the sea. Northumberland County Council led a campaign against the proposed power station but for almost 20 years the scheme was kept as an option by the government and nuclear industry. Only after privatisation in 1996 was it announced that there were no plans to build any more nuclear power stations in the UK - at least not in the foreseeable future.

The coastal system at Druridge Bay has survived many pressures but not without some damage to the environment. It remains a beautiful beach, popular with tourists, and an important wildlife site. However, there is also erosion of the beach and dunes, and the presence of nearby open cast coal workings.

Take the role of a planning consultant working for the County Council. Limited European Union funding has become available for improving damaged coastal environments. The Council wants to submit a bid for the funding but must choose between three options:

1 Buy the former nuclear power station site and develop into a large nature reserve.
2 Develop new amenities for leisure and tourism in the Country Park.
3 Spend the money on coastal protection schemes to reverse the erosion that has taken place.

You are asked to write a short report. It should:
- outline the advantages and disadvantages of each option
- include a sketch map of the Bay
- advise which option, in your view, would be the best use of the European funds.

Sand dunes are very fragile environments. Trampling by people kills the vegetation so allowing the sand to blow inland.

Glossary

All the terms listed below are explained in this Enquiry. In each case, write a definition of the term and, if possible, give at least one example.

Physical system	Transport (of eroded material in a river or along a coast)	Catchment area
Input		Watershed
Physical process	Traction	Source
Output	Saltation	Tributary
Hydrological Cycle	Suspension	Braided channel
Surface run-off	Solution	Wave cut platform
Infiltration	Deposition	Cliff
Groundwater	River cliff	Stack
Transpiration	Delta	Arch
Throughflow	Interlocking spurs	Longshore drift
Water table	Flood plain	Groyne
Erosion	Terrace	Bar
Hydraulic action	Meander	Spit
Corrasion	Slip-off slope	Tombolo
Attrition	Drainage basin	

Coursework Enquiry

River Study

Hypothesis: The character of a river changes as it flows downstream.

This is a popular coursework topic because it is relatively easy to carry out. It nevertheless provides a good opportunity for first hand study of physical processes and landforms.

You need to choose at least 3 sites along a stream or river where you can gain access. At each site you have to wade safely across so it is most likely that the best locations will be nearer the source rather than the mouth of the river.

At each site you will collect data on the following:
- channel width and depth; river velocity.

Additionally you might investigate the bedload (size and shape) together with the physical and human environment at each site.

Method of enquiry and report writing

1 Explain your hypothesis - why is the hypothesis likely to be true? What does it say in this and other books about rivers and their landforms downstream from the source?

2 Decide what data you need - you have to consider what information you need from each site. Bear in mind the practicalities of selecting and getting to each site. There is no point in deciding what data you need before making sure that you can gain access to suitable, safe sites.

3 Data collection - you will need an OS map showing your river but most of your research will be by observation and measurement. Field sketches and photographs provide useful records of each site. You will need some equipment for this enquiry: a measuring tape, metre rules (or poles with measurements marked), a watch (which measures seconds) and floats (oranges are often used because they float partially under water and are not so affected by the wind). (Wear either wellington boots or trainers that you do not mind getting wet.)
Plan the data collection in advance and prepare recording sheets for each site.

4 Collect the data - for this enquiry you must work with at least two other people. Measure the width by holding or attaching the measuring tape at the water's edge on each bank. Keeping the tape in position, measure the depth at regular intervals across (say at ten different points).

Then mark out a 10 metre section of the stream and carefully time the float between the points you have marked. Do this several times and calculate the average time. If the river is wide, time the float at each side of the channel as well as in the middle. (Your school might have a flow vane or flow meter which would provide more accurate results.)

For the bedload, measure the long axis of 50 randomly selected pebbles or rocks on the stream bed. The 'big toe' method gives reasonably random results - simply walk a pace and measure the pebble or rock in front of your toe. Your teacher may have a roundness chart for you to use, otherwise devise your own 5 point scale from very rounded to very angular.

Draw field sketches or take photographs to record the physical and human features around your chosen sites.

5 Data presentation - draw maps, graphs and diagrams to display your results. You can calculate the channel cross-sectional area by multiplying the average depth by the width. The velocity is calculated by dividing the distance travelled (10 metres) by the average time in seconds. (Multiplying by 0.85 'corrects' your result because surface water velocity is slower than within the river.) Discharge (m^3/sec) is calculated by multiplying velocity by cross sectional area.

6 Describe and analyse the data - say in words what your graphs and diagrams show.

7 Accept or reject the hypothesis - do your results support or contradict the hypothesis? To what extent, if any, do river width, depth, velocity, bedload size and shape, physical and human environment vary with distance downstream? Try to explain your results by referring to river processes.

8 Follow up - did everything in your enquiry go to plan? Which aspects were more successful than others, and why? What external factors (such as the weather and recent rainfall) might have affected your results? With more sophisticated measuring instruments, how might your study be improved?

As a follow up you might carry out the same study at a different (wetter or drier) time of year and compare your results.

9 Global environments and ecosystems

to the student

The world can broadly be divided into a number of *ecosystems*. These are zones of natural vegetation and associated wildlife. They interact with their physical environment. For example, a rain forest is created by physical conditions such as climate and soil. At the same time, the climate and soil are affected by the vegetation in the rain forest.

In some parts of the world, much of the natural vegetation has been cleared for farming or urban development. For instance, in the United Kingdom much of the country was once covered by deciduous woodland. Today, woodland covers less than 10 per cent of the whole country.

In many areas of the world, natural vegetation is under threat from farming and other human activity. This Enquiry looks at various ecosystems and considers how human activity is affecting them.

questions to consider

1 Where are the main global ecosystems located?
2 What are the vegetation and physical features of different ecosystems?
3 How does human activity affect particular ecosystems?
4 How can people protect and sustain fragile environments?

key ideas

Before human activity the land surface of the world was covered by zones of *natural vegetation*. In some places, this was grassland, in others it was forest or woodland. These zones were influenced by the type of climate, soil and landscape found there.

The way that plants and animals (the *biological environment*) interact with climate, soil and landscape (the *physical environment*) is described as an *Ecosystem*.

Ecosystems are often altered by human activity. Sometimes the changes are very destructive. Human activities which use and restore natural resources rather than destroy them are known as *sustainable development*.

activities

Using information from this and the facing page:

a) Give five examples of global ecosystems.
b) List three physical factors that affect natural vegetation.
c) Working with a partner, 'brainstorm' a list of human activities that affect ecosystems.

Climate

Soils

Landscapes

Human activity

Coniferous Forest

Desert

Rain Forest

Savanna

Grassland

enquiry

Rain forests and the tundra - two ecosystems under threat

In this Enquiry you will investigate ecosystems - what they are, where they are located and how they are affected by human activity. Some fragile environments and ecosystems are particularly threatened. The pages that follow concentrate on two examples - the rain forest and tundra ecosystems.

You will be asked to consider the threats to these particular environments and the ways in which people try to protect and conserve them.

The outcome of the Enquiry will be an article which you will be asked to write. The article will focus on a particular threatened environment - either in a rain forest or tundra region. You will be asked to suggest strategies that can be used to overcome the problems that arise in these ecosystems.

STAGE 1: Where are global ecosystems?

The world can be divided into a number of different ecosystems. These broadly match the world's main climatic zones. For example, rain forests are found in zones which have a hot and wet climate throughout the year.

Today, human activity has cleared or changed natural vegetation in all countries of the world. Only small pockets of completely unchanged forest or grassland remain. Nevertheless, it is usual to show on a world map the location of ecosystems before human activity.

Using the resources which follow, classify the global ecosystems into:

a) tropical (ie, situated between the Tropic of Cancer and Tropic of Capricorn)

b) temperate (ie, situated between the Tropics and Arctic / Antarctic Circles)

c) polar (ie, situated between the Arctic / Antarctic Circles and the Poles).

Make a note of those ecosystems which 'overlap' and occur in, say, both tropical and temperate regions.

Then, use an atlas to complete a table like the one shown below. For each ecosystem shown on the page opposite, list some countries where the ecosystem can be found and describe the climate.

Human activity has altered most natural vegetation. This coniferous forest has been clear cut in the north west USA.

Ecosystem	Countries	Climate summary
Tropical rain forest	Brazil, Zaire, Indonesia	Hot and wet throughout the year

On the page below, different climates are described. The following table is designed to help you understand what is meant by terms such as 'hot' or 'mild',

very hot	**> 30 °C**
hot	**20 - 30 °C**
warm	**15 - 20 °C**
mild	**10 - 15 °C**
cool	**0 - 10 °C**
cold	**< 0 °C**

Tundra: very cold long winters. A short (2 or 3 months) summer when the average temperature rises above freezing. Plants have a very short growing season.

Coniferous forest (called taiga in Russia): cold for most of the year. Average temperature rises above freezing for 5 or 6 months of the year. The trees have pine needles which allow them to conserve water while the ground is frozen.

Deciduous woodland (the UK is located in this zone): cool winters and warm summers with rain throughout the year. Wet enough for trees to grow. In winter the trees stop growing and shed their leaves.

Temperate grassland (called prairie in North America): cold winters and hot summers with a summer rainfall maximum. Too dry for trees to grow.

Rain forest: hot and wet for all or most of the year. The climate creates ideal growing conditions for trees and other plants.

Tropical grassland (called savanna in Africa): very hot wet summers, hot dry winters. High rates of evaporation create conditions too dry for most trees. Fires (man made or natural from lightning) also reduce tree cover.

Desert: dry throughout the year. Deserts can be in hot or cold regions. Plants must be able to withstand drought.

Mediterranean type scrub and evergreen woodland (found around the Mediterranean Sea and in similar latitudes on the western margins of continents). Hot dry summers and mild, wet winters. The plants have to withstand the summer drought.

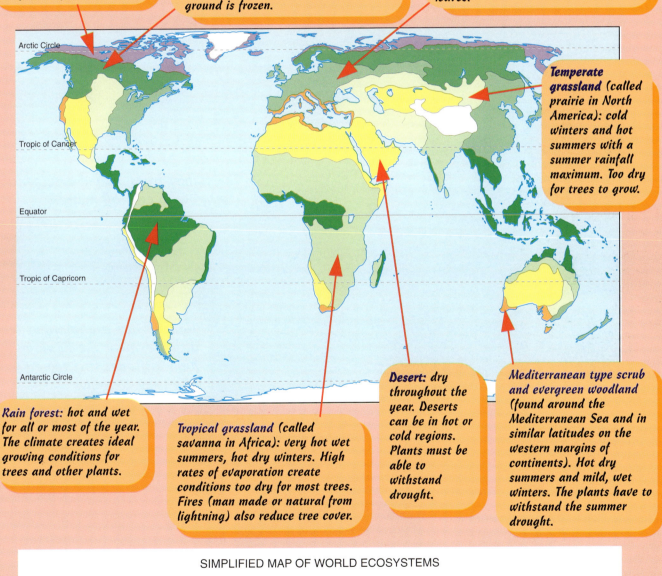

Arctic Circle
Tropic of Cancer
Equator
Tropic of Capricorn
Antarctic Circle

SIMPLIFIED MAP OF WORLD ECOSYSTEMS

- Rainforest
- Tropical grassland (savanna)
- Temperate grassland
- Desert
- Mediterranean type scrub and evergreen woodland
- Deciduous woodland
- Coniferous forest (taiga)
- Tundra
- Ice cap and mountains

STAGE 2: Two ecosystems compared

Natural vegetation is a term that describes the plants that grow in an area if there is no human activity. In the resources that follow, two zones of natural vegetation are described. In addition, the factors such as soil which affect, and are affected by, the plants are described and explained. These interacting factors - biological and physical - form a complete ecosystem.

Use the resources to produce a table. It should have row and column headings like this:

	Tropical rain forest	Tundra
Description of vegetation		
Climate		
Soil		
Landscape features		

Tropical rain forest

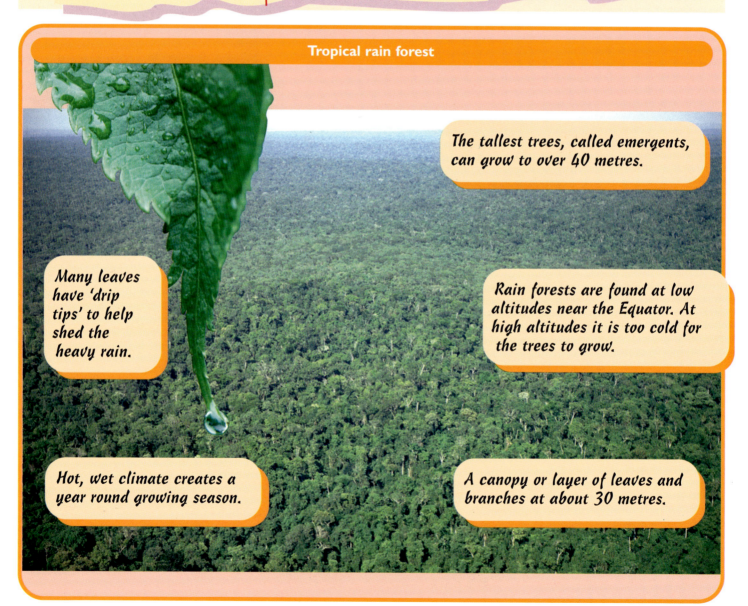

The tallest trees, called emergents, can grow to over 40 metres.

Many leaves have 'drip tips' to help shed the heavy rain.

Rain forests are found at low altitudes near the Equator. At high altitudes it is too cold for the trees to grow.

Hot, wet climate creates a year round growing season.

A canopy or layer of leaves and branches at about 30 metres.

Because trees lose their leaves at different times, the forest is always green.

Long, straight tree trunks - the trees grow quickly to reach the light.

Large variety of different species scattered throughout the forest.

Conditions in a rain forest are like those found in a hot, humid greenhouse. The plants grow straight and fast to compete for sunlight. Nutrients from rotting leaves and vegetation are quickly recycled and absorbed by tree roots.

Buttress roots support the massive tree trunks.

Dark forest floor with little undergrowth (except in clearings or on a river bank).

Layer of quickly rotting leaves creating a fertile soil.

Extends across Alaska, northern Canada, Scandinavia and Russia. Also found at high altitudes in mountain areas throughout the world.

Only 2 or 3 months with an average temperature above freezing.

Windy conditions are common.

Complete snow cover for at least 6 months.

Low amount of precipitation but little evaporation.

In winter, 2 or 3 months of permanent darkness. In summer, 2 or 3 months of permanent daylight.

Conditions on the Arctic tundra vary from extreme cold and permanent darkness in winter to cool, permanent light in summer. During the brief growing season plants flower and seed quickly. The ground is boggy and waterlogged when it is not frozen.

Tundra conditions exist in most high mountain regions from the Scottish Cairngorms to the Himalayas.

Growing season is too short for trees to grow.

Thin layer of peat, organic matter decomposes very slowly.

For a short period in mid summer the tundra blooms with wild flowers.

Permanently frozen subsoil (permafrost), waterlogged upper layer in summer.

Low lying, dwarf plants, lichens and mosses.

STAGE 3: How does human activity affect particular ecosystems?

Ecosystems can be affected by natural events such as volcanic eruptions or fires caused by lightning. However, the most important factor that affects ecosystems is human activity. This might be farming, quarrying, mining, forestry, road building or urban development.

The resources that follow describe human activity in the two ecosystems described in Stage 2. They show how different activities, such as oil exploration in the tundra and forestry in the rain forest, affect the ecosystem. Draw line sketches of a rainforest and also a tundra scene. On the sketches list the human activities that are affecting these ecosystems. Then, underneath, briefly describe the effect of these activities.

Rain forests

Tropical rain forests provide hardwoods such as mahogany, ebony and teak. These woods are strong and resistant to rotting so are in high demand for constructing buildings and furniture. The forests contain a great diversity of plant and animal species (this is known as **biodiversity**). Papua New Guinea, to the north of Australia, has one of the largest rain forest wildernesses remaining on earth. It contains over:

- 200 mammal species
- 700 bird species
- 11,000 plant species (including 1,500 tree species)
- 700 indigenous (ie, local and native to the area) human cultures.

Countries where rain forests are threatened

1 Costa Rica	7 Ghana	13 Thailand
2 Venezuela	8 Togo	14 Malaysia
3 Brazil	9 Benin	15 Philippines
4 Senegal	10 Nigeria	16 Indonesia
5 Liberia	11 Cameroon	17 Papua New Guinea
6 Ivory Coast	12 Zaire	

Why are rain forests under threat?

- Farmers and settlers clear the forest in order to grow food. In remote areas of low population density, the people might '**slash and burn**'. This means that they cut down and burn the trees in order to grow crops for 2 or 3 years. They then move on to another area allowing the forest to regrow and regenerate.

  However, in areas where population pressure is greater - such as parts of Amazonia in Brazil, the farming might be more permanent. Alternatively, the forest might be cleared for cattle ranching.

  Problems created: once the forest is cleared, the soil quickly loses its fertility because there is no longer a layer of rotting vegetation being recycled. Without the layer of protective vegetation, the heavy rain washes out the soil's nutrients and causes rapid soil erosion.

- In some rain forests the trees have been cleared for plantations. Examples include rubber plantations in Liberia, West Africa and palm oil in Papua New Guinea, South East Asia. This land use has the advantage of keeping a vegetation cover and so protecting the soil from erosion. It also provides income and employment for local people.

  Problems created: the products from plantations are sold to developed countries on world markets and sometimes their price falls. Because the land is not used for food production, people have to buy their food from elsewhere. If their income falls, they have no means of subsistence. Also, **monoculture** (ie growing of a single plant species) robs the environment of its biodiversity and is at great risk from pests and disease.

A coconut plantation. This provides income for local people and protects the soil from erosion. However, monoculture can bring risks.

- Because tropical hardwoods are so valuable, there is a big incentive for multinational logging companies to cut the forests down. When this is unregulated, the cheapest method is to bulldoze tracks into the forest and '**clear cut**' all the trees (ie, cut everything down) and then remove those logs that are most valuable. This provides a 'one off' source of income to the landowners who sell the logging rights.

  Problems created: the bulldozed roads and cleared forests suffer from rapid soil erosion - making it hard for new growth to reestablish itself. The soil erosion causes streams and rivers to become silted and unusable as a water supply. On the coast, corals die as they become covered in river sediment. Once the logs have been exported, the logging company moves on to another area or country - having destroyed the environment for the local people and leaving them with no means of survival.

A rain forest being cleared for cattle ranching. The bare soil is vulnerable to erosion during rainstorms.

Why is the destruction of rain forests seen as a world problem?

- Many people believe that it is morally wrong to destroy the culture and means of subsistence of the indigenous people who live in the forests and who cannot protect themselves.

- The rain forests are the richest and most biodiverse ecosystem on earth. By clearing them, we are losing hundreds and even thousands of species of plants and animals. Once a species is extinct, it is gone forever. Because medicinal drugs have sometimes been discovered and extracted from plants, this loss might make it harder to find cures for diseases.

- Rain forests are sometimes called the 'lungs of the planet'. This is because they absorb so much carbon dioxide from the atmosphere (and then release oxygen). By storing carbon dioxide, a 'greenhouse gas' - the danger of global warming (see page 118) is reduced. [Note: a greenhouse gas acts like a pane of glass in a greenhouse. It traps solar radiation and so causes the temperature on the earth's surface to rise. Some people believe that if current trends continue, rising temperatures will cause some of the Arctic and Antarctic ice caps to melt. If this happens, many low lying coastal areas will be flooded. At the same time, the rising temperature will create unpredictable climatic changes - making some places drier and some places wetter.]

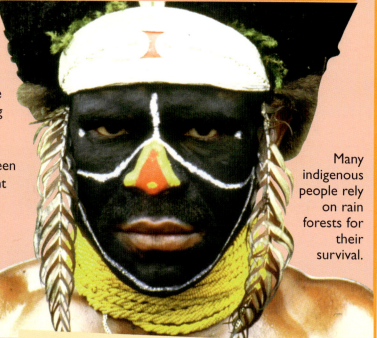

Many indigenous people rely on rain forests for their survival.

Rain forest destruction in the 1980s (selected countries)

	Forest total area (000 sq km) 1990	% change 1981 - 1990
Brazil	5,611	- 6
Zaire	1,133	-6
Indonesia	1,095	-10
Venezuela	457	-12
Papua New Guinea	360	-3
Cameroon	204	-6
Malaysia	176	-20
Thailand	127	-33
Philippines	78	-33
Costa Rica	14	-29

When rain forests are destroyed, the danger of global warming increases.

Some quotes on the rain forest

"Our people live in the forests. They do not have shops. They depend on the forest for their meat, streams to fish and plants to eat. Once the big companies start logging the area, my people will lose everything." **- Theresa Morupe Haihue, woman from Sandaun province, Papua New Guinea.**

"We have no water to wash in and hardly any water to drink that is not polluted. Our secret places and wild animals are gone, and all our ancient trees are being destroyed." **- Yalaum Mosol, Bemal village, Papua New Guinea (site of clear cut logging by a multinational timber company).**

"About 80% of Papua New Guinea (PNG) is covered by rain forest, one of the world's outstanding resources. The big logging companies have cleared much of the accessible forest in Malaysia, Thailand, Cambodia and the Philippines. Now, industrial logging has increased in PNG by 400 per cent in one year. There is virtually no reforestation and the trees are being cut down at three times the sustainable rate." **- report in PNG newspaper.**

"In the 1980s, a tenth of the world's tropical rain forest was destroyed - an area the size of France, Germany, Italy and the UK put together. In the 1990s, the pace has quickened. In Africa and Asia, it is loggers who are doing most of the clearing. In Latin America, it is farmers - including those growing illegal drugs such as cocaine." **- report by Food and Agriculture Organisation of the United Nations.**

"The accessible coastal forests of West Africa, in Ghana, Cote d'Ivoire, Nigeria, Togo, Senegal and Benin have already been cleared of their best timber and have lost many species of animals. Boom has turned to bust and the forest to scrub. Now the plunder has shifted to Central Africa - in Cameroon, Zaire, Gabon and the Central African Republic. Cameroon alone is losing 200,000 hectares a year." **- Claude Martin, director general of World Wide Fund For Nature (WWF).**

Logging companies in South East Asia are cutting down forests faster than they can grow.

Logging is dominated by large multinational companies.

Arctic tundra

The tundra is one of the world's last great wildernesses. It extends through Alaska (USA), Canada, Greenland (Denmark), Iceland, Norway, Sweden, Finland and Russia. It is home to scattered groups of Inuit (Eskimo) and other native hunter gatherers. Wildlife includes polar bears, grizzly bears, foxes, musk ox and deer (such as caribou and reindeer). In summer, millions of migratory birds fly north and make nests in the tundra region.

Tundra is one of the world's most fragile environments. The harsh climate leaves little flexibility for survival.

Because the growing season is so short, species that are harmed have little time for regeneration. The cold temperatures and low amount of solar radiation slow the breakdown of organic matter - including contaminants such as oil. As a result, once in the Arctic food chain, these contaminants tend to endure longer than they do in southern latitudes.

Caribou on the Alaskan tundra.

Why are tundra regions under threat?

- minerals such as oil, gas, coal and precious metals are all found in tundra regions. At Prudhoe Bay, on the north coast of Alaska, the largest oil field in the USA was discovered in 1969. Since then, exploration has quickened. In the Yamal Peninsula in north west Siberia (in Russia), a massive gas field has been discovered. Roads, pipelines and buildings to house the workers are all required when mineral deposits are developed.

- during the Cold War between the USA and the USSR, the enemies faced each other across the Arctic. Early warning radar stations and military bases were built in the tundra. In the Murmansk area of northern Russia the rusting remains of a nuclear submarine base threatens radioactive pollution across the whole northern hemisphere.

- the Canadian Minister of the Environment called the Arctic 'an early warning system for our planet'. Wind blown pollution from factories in Asia, Europe and America are spreading poisons in the Arctic food chain. High levels of PCBs (a pollutant known to cause cancer), DDT (a dangerous pesticide) and heavy metals such as mercury and cadmium have been found in foods eaten by people in Canada's Mackenzie River Delta.

- pollutants such as CFCs (from refrigerators, air conditioning systems and aerosols) have caused ozone levels in the upper atmosphere to fall. Each spring, a 'hole' in the ozone layer now appears over Arctic regions. Ozone has the effect of reducing ultraviolet radiation from the sun so the 'hole' allows in more radiation. This is dangerous to humans in Arctic regions because it causes skin cancer, eye cataracts and damage to the body's immune system.

An area under threat - the Arctic National Wildlife Refuge (ANWR)

This is a 10 million hectare refuge on the north coast of Alaska, near to the Canadian border. It is the scene of a major dispute in the USA between conservationists and oil developers. The Refuge lies 120 miles to the east of the Prudhoe Bay oil field and holds estimated oil reserves of 9 billion barrels - making it the second biggest oil field in the USA after Prudhoe Bay.

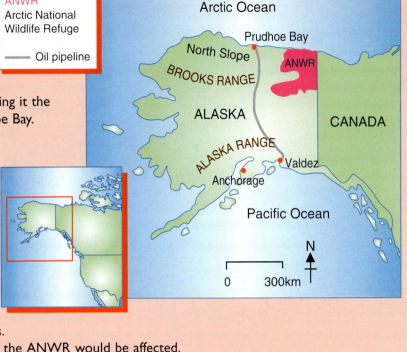

ANWR
Arctic National
Wildlife Refuge

—— Oil pipeline

The case for oil development:

- production at Prudhoe Bay has peaked and is now in decline. ANWR is the USA's best hope for a new domestic oil supply.
- domestic oil production reduces the USA's dependence on imports of foreign oil from politically unstable regions (such as the Persian Gulf).
- the oil production would increase the USA's Gross National Product by $50 billion and would create thousands of jobs.
- less than 1 per cent of the coastal plain in the ANWR would be affected.
- Prudhoe Bay shows that development can take place without damaging the environment.
- local people have benefited from Prudhoe Bay with jobs, new homes, street lights, power plants and medical clinics.
- the state of Alaska receives 85 per cent of its tax revenue from the oil industry.

The case against oil development:

- the ANWR is the USA's last great wilderness. It is sometimes called the 'American Serengeti' (a famous animal reserve in East Africa) because it contains such a diverse wildlife.
- the migration route and calving area of a 160,000 strong caribou herd would be threatened.
- the Gwich'in people (the local native Americans) would risk losing their traditional culture of following and hunting the caribou herd. The danger is that if they change their way of life, they will suffer the same problems of similar disturbed communities - unemployment, living on welfare payments and alcohol abuse.
- oil exploration and production, including drilling, seismic surveys, explosions, road construction, toxic waste pits, garbage dumps and traffic from bulldozers, snowmobiles, helicopters and planes would drive away the polar bears and other animals.
- an oil spill would permanently destroy the fragile vegetation. Oil remains toxic for long periods in the Arctic because it gets trapped in the ice and only breaks down slowly in the cold temperatures.
- over 90 per cent of Alaska's Arctic coastline is already open for oil exploration. The remaining 10 per cent in the ANWR should be protected.
- oil is a non renewable resource and is being wasted in the USA. If the average 27 mpg of American cars could be raised to 40 mpg, the oil from the new field would not be required.

Oil decomposes very slowly in the cold Artic environment.

169

STAGE 4: How can people protect fragile environments?

A **fragile environment** is one which is easily damaged by human activity. In the UK, many people now believe that it is important to protect the environment. By-pass protesters sit in trees to prevent them being cleared for new roads. Shops like B&Q only stock timber goods from forests which are replanted after cutting. Pressure groups like Greenpeace campaign against companies and governments who, it believes, are damaging the environment.

This country has almost no completely natural vegetation left. Woodlands have been cleared for farming and even upland 'natural' landscapes have been changed by grazing or burning to improve pasture. Nevertheless, traditional landscapes of hedgerows and open moorland - even though they are not completely natural - are fiercely protected by **conservationists**.

Similar issues are important in other countries. In some cases, though, they are more matters of life and death. For example, **desertification** (the spread of deserts) in parts of Africa, threatens the way of life of farmers and herders. Without pasture for their cattle and goats, they lose their livelihoods and means of survival.

Examples of people trying to protect their environments are given in the resources that follow.

On an outline world map, locate and mark the examples described in the resources. In each case, make a note that describes and explains:

- how and why the environment is threatened
- the ways in which people are trying to protect the environment
- the case, if any, for exploiting the natural resources
  (also use information from Stage 3 to help you in these tasks).

Rain forests

Save our Planet Incorporated

Mission Statement: Helping the people of today preserve our world for the people of tomorrow.

Save our Planet is a company operating in Costa Rica. It says that the notion of leaving rain forests untouched is old fashioned and unrealistic. The new thinking, it says, is that the forests are factories and must be used to offer local people an alternative to clearing for farming. This is because, once the forest is cut and burned, the soil can only grow good crops for two years at most. So, since 'slash and burn' is not acceptable, some way must be found to let local people make a living. This idea is called **sustainable development**. It involves managing the forest rather than destroying it. The forest is surveyed and a calculation is made on how much timber can be 'harvested' without damaging the ecosystem. Trees are catalogued on a database and given a specific year when they will be mature and ready for cutting. Helicopter extraction is suggested as a way of avoiding bulldozer tracks. The system avoids 'clear cutting' whole areas and allows time for the trees to grow - so the forest as a whole survives.

Sustainable development involves renewing and replanting the forest, rather than destroying it.

The Forest Stewardship Council (FSC)

This international organisation was set up in 1993. It gives a seal of approval to timber and timber products that come from well managed, sustainable forests. When the forests have been 'certified' (ie, given the seal of approval), the timber can be stamped with the FSC logo. The system allows shops and consumers to know that the products they buy are from 'eco-friendly' producers. This means that the forests are not clear cut. Instead, individual, selected trees are cut down, new trees are planted and wildlife habitats are preserved. At the same time, the economies of the producer countries benefit by the export of timber and local people benefit by having jobs in the timber trade.

In the UK, the scheme has been backed by the World Wide Fund For Nature (WWF). By a mixture of publicity and consumer pressure, the WWF has persuaded retailers like B&Q, Boots, Tesco and Sainsbury to phase out any timber products that do not have the FSC seal of approval.

The first areas of tropical rain forest to be certified are in Malaysia, Papua New Guinea and Cameroon.

Mahogany is murder: a statement from Friends of the Earth

Every time we buy a piece of Brazilian mahogany, we help to fund the destruction of the Amazon rainforest and the people who live there.

Britain is the second largest importer of Brazilian mahogany. 80 per cent of mahogany sold here is logged in Amazonia. Most of this mahogany is stolen - illegally cut in Indian reserves - by gangs who plunder the forests, bringing disease and destruction with them. Those who get in their way are driven off the land and even shot. Tropical rain forests are the richest source of life on earth, home to over half of the world's species of plants and animals. Time is running out, forests are disappearing, people are dying. Please help us to stamp out this murderous trade once and forever.

What you can do:
  - don't buy mahogany
  - write to the Prime Minister urging a ban on the import of Brazilian mahogany.

Extracts from a speech by a pro-logging government minister in Papua New Guinea

The government needs money for basic services, schools, clinics, bridges and roads. The taxes we get from the logging companies can pay for these services. Most of the so called advanced world and newly industrialising countries have built their prosperity on the wholesale exploitation of the very resources we are told to protect. We do not like being told by other countries not to do the very things they did to get ahead. If the world expects us to preserve a resource which is valuable to all the world, then the cost of our restraint must be shared with the world.

Look at the basic statistics for our economy. You will see how much we need the money from the logging companies.

Papua New Guinea

GNP per capita (1994, US$): 1,240		(UK: $18,340)
Life expectancy:	57	(UK:76)
Adult literacy (%):	28	(UK:>95)
Population with access to safe water (%):	33	(UK:100)
Infant mortality rate (per 1,000 live births):	65	(UK:6)

BP in Prudhoe Bay

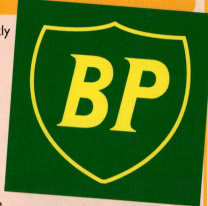

When oil was discovered on Alaska's North Slope at Prudhoe Bay, in 1969, it quickly became clear that this was the biggest oil field in the USA. It was an important discovery because other US oil fields were past their peak and the country was becoming increasingly reliant on imported oil. However, conservationists were immediately concerned that the natural environment of 'America's last wilderness' would be threatened. BP therefore made a huge effort to reassure people that the development could be achieved without lasting damage to the environment. For example, caribou herds were tracked using radio transmitters and helicopters to make sure that their migration routes would remain clear of development.

BP wished to protect the Arctic tundra environment for a number of reasons. If the company had destroyed large natural areas it would have found it difficult in the future to get permission for more oil exploration. Secondly, big multinational companies realise that public opinion is important, as Shell found out when it tried to dump the Brent Spar oil platform in the North Atlantic. Pressure groups campaigned against the company and, in some countries, urged a boycott of Shell petrol. Therefore, public relations and public image are important for sound commercial reasons. Thirdly, any environmental damage might have resulted in a legal action by the Alaskan government. When the Esso tanker, Exxon Valdez, spilled oil at Valdez in 1989, the company was ordered to pay $5 billion in damages. Since then, the USA has passed a law, the Oil Pollution Act, that allows unlimited liability for oil spills. It means that a company can be bankrupted if found responsible for an oil spill.

Prudhoe Bay oil well.

In addition to building the oil wells on the North Slope, BP had to build an 800 mile long pipeline across Alaska to an ice free terminal at Valdez. This was completed in 1977 after two years work using up to 20,000 workers at the peak. Now, up to 52,000 gallons of oil per minute can be pumped to Valdez. In past gold rushes, towns had been built in Alaska's wilderness and then abandoned. BP made sure that the housing for the construction workers was dismantled and the landscape restored after the pipeline was finished.

Valdez oil terminal.

Alaska oil pipeline.

The pipeline and housing is built on stilts because, otherwise, their heat would melt the permafrost and cause the foundations to sink. To avoid damaging oil spills, the pipeline was built to very high specifications, coated and then protected by an insulating jacket. Helicopters with infra red cameras patrol the pipeline to detect any heat loss (oil from underground is at a high temperature).

This heat loss might be from a leak. Alongside the oil wells and pipeline, gravel roads have been built to allow road transport. Normally, this is impossible in summer because the waterlogged ground cannot cope with the weight of a vehicle. BP trialled and tested varieties of grass and other plant species that would survive the severe conditions in order to cover over the workings that were necessary during construction.

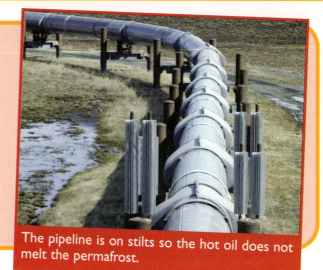

The pipeline is on stilts so the hot oil does not melt the permafrost.

STAGE 5: Review

You are asked to select one threatened environment - either in the rain forest or the tundra. Take the role of a journalist researching and writing an article for a newspaper or magazine. You should include maps and diagrams where appropriate. Your task is to write an article that:

- describes the ecosystem
- describes and explains the factors that have helped create the ecosystem
- describes the threat to the area
- suggests strategies that people (eg local people, companies, governments, international bodies) are using and might use in the future to protect the environment
- use headlines and, if possible, illustrations to emphasise the main points.

Glossary

All the terms listed below are explained in this Enquiry. In each case, write a definition of the term and, if possible, give at least one example.

Ecosystem	**Slash and burn**
Natural vegetation	**Monoculture**
Biological environment	**Clear cut**
Physical environment	**Greenhouse gas**
Sustainable development	**Fragile environment**
Permafrost	**Conservationist**
Biodiversity	**Desertification**

10 Industrial decline and change

to the student

In the last half of the twentieth century there have been huge changes in the types of industry and the jobs that people do in Britain, Europe and North America. There has been a massive decline in the number of people working in manufacturing. Hardest hit have been men, especially those employed in *heavy industries*. These include mining, iron and steel, and heavy engineering such as shipbuilding and locomotive manufacture. At the same time there has been a large increase in the numbers employed in *service industries* (eg, finance, health care, tourism and transport).

In this Enquiry, we shall look at the causes and effects of these changes in two major industrial areas, the North East of England and the Ruhr in Germany. Both regions were at one time dominated by heavy industry but are now struggling to change and adapt to new circumstances.

questions to consider

1 What were the traditional industries of Britain and Europe in the first half of the twentieth century?
2 Why did these industries decline?
3 What effect did this decline have?
4 How have the old industrial regions changed?

key ideas

Industries and occupations can be classified into *primary*, *secondary* and *tertiary*.
Changes in technology and the pattern of demand, together with new competition from lower cost producers, have led to decline in some industries.
Older industrial areas have suffered from higher unemployment than those relying on light manufacturing and services.
In Europe and North America, employment in the primary and secondary sectors has declined and there has been an increase in the numbers employed in the tertiary sector.

activities

Using information from the facing page:

a) Suggest reasons why Britain was the world's leading shipbuilder in the early years of the twentieth century.
b) Now most of the world's ships are made in Taiwan and South Korea. Why? Make a list of possible reasons why British shipbuilding has declined.

INDUSTRIAL DECLINE AND CHANGE

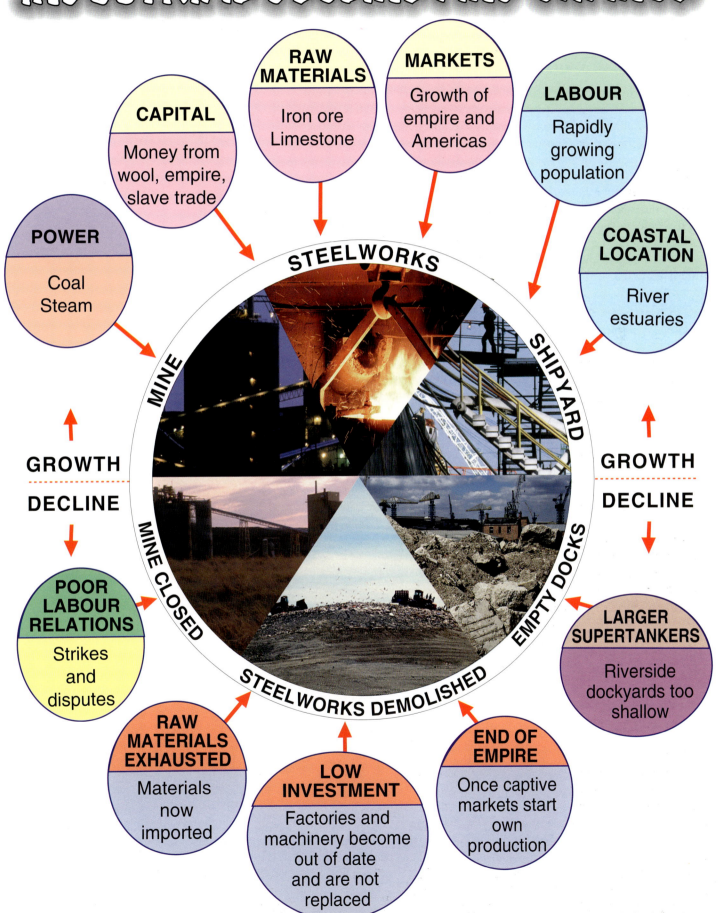

CAPITAL
Money from wool, empire, slave trade

RAW MATERIALS
Iron ore Limestone

MARKETS
Growth of empire and Americas

LABOUR
Rapidly growing population

POWER
Coal Steam

COASTAL LOCATION
River estuaries

STEELWORKS

MINE

SHIPYARD

GROWTH

DECLINE

GROWTH

DECLINE

MINE CLOSED

EMPTY DOCKS

STEELWORKS DEMOLISHED

POOR LABOUR RELATIONS
Strikes and disputes

RAW MATERIALS EXHAUSTED
Materials now imported

LOW INVESTMENT
Factories and machinery become out of date and are not replaced

END OF EMPIRE
Once captive markets start own production

LARGER SUPERTANKERS
Riverside dockyards too shallow

The changing world of work

In the USA, the regions in the north east of the country that suffer from declining industries are called the 'rustbelt'. This is because they contain derelict and abandoned factories, steel works and mines. The industries have closed down or moved on, leaving behind populations which have to face the social and economic problems of high unemployment.

In Europe the same process has been taking place. The graph shows what has happened to the coal mining industries of Germany and Britain over the past 50 years. The outcome of this Enquiry will be an essay that you will be asked to write. It will focus on the changes that have taken place in one of the industrial areas studied in this Enquiry - ie, North East England or the Ruhr in Germany. The various Stages of the Enquiry will provide you with the information, maps and statistics that you will need.

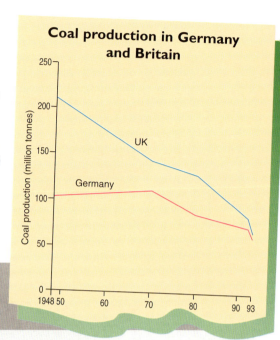

Coal production in Germany and Britain

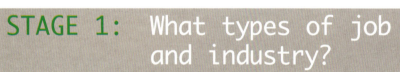
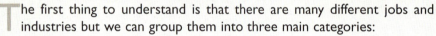

The first thing to understand is that there are many different jobs and industries but we can group them into three main categories:
- primary
- secondary
- tertiary.

In the resources that follow, you will find definitions, job adverts, examples of each type of industry and statistics to show the pattern of employment in different areas.

Look at the job advertisements that are displayed opposite. Write them in a list and, in each case, decide whether they are in the primary, secondary or tertiary sector.

Consider the triangular employment graph that shows a comparison between different countries. Make a brief written note that compares the employment patterns (or **occupational structures**) in countries with different levels of income? Then, using the information provided, mark on a triangular graph the employment pattern of North East England and North Rhine - Westphalia, the German region that contains the Ruhr. Write a short description of what the triangular graph shows. Compare the UK and German regions with the other EU regions.

Sometimes, expanding industries based on new technology are labelled 'sunrise'. This is because they are associated with the east, ie Japan and East Asia. Declining industries have come to be labelled 'sunset'.

Job advertisements

SHEPHERD

Wanted for a flock of 250 pedigree Texel Ewes
Skill needed to prepare sheep for shows, sales and carcass competitions
Good salary and House provided.

Computer Officer (Multimedia Development)

Information Services
Applicants are invited to work in a team providing assistance to users of PC based systems.

COMMERCIAL VEHICLE MOTOR BODY BUILDERS

Required, skilled, semi-skilled and trainees.
Due to expansion we need extra staff.
Must be well motivated.

Due to expansion

MAYTEX CURTAINS

require

MACHINISTS AND CUTTERS

Good rates of pay and working conditions.

Urgently Required

JAPANESE TRANSLATOR / TYPESETTER

Must have 5 years experience and speak and write Japanese fluently.

ASSISTANT TO SECRETARY

of small charity

This is a full time post. Applicants should have administrative experience and be computer literate. A knowledge of accounts an advantage.

Trained **FITTERS** required for work on off-shore gas platform

Must be skilled and ready to work long hours

Four weeks on and four weeks off

Excellent pay and fringe benefits.

EXPERIENCED DECKHANDS REQUIRED

Scottish trawler needs two experienced fishermen.

Generous bonus scheme.

What type of job?

Although there are many different types of jobs, we can group them into three main classifications:

1 **Primary sector** - these jobs involve collecting or extracting material provided by nature, for example, farming, forestry, fishing, quarrying and mining. So, a dairy farmer, a worker on a sugar cane plantation, a coal miner and an oil rig operator all work in the primary sector. They provide the **raw materials** for the secondary sector.

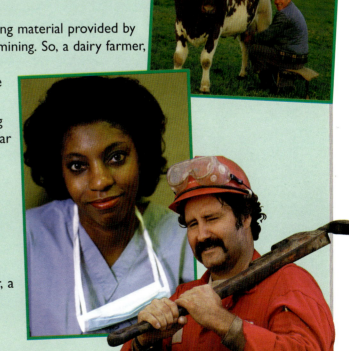

2 **Secondary sector** - these jobs involve processing, making or manufacturing a product. For example, a steel worker, a car assembly line worker, a textile machinist and a worker in a bakery or brewery are all employed in the secondary sector. Steel, cars, textiles, loaves of bread and cans of beer are all **manufactured goods**.

3 **Tertiary sector** - these jobs provide **services** for the other two sectors and also for the wider community, for example, in transport, communications, health, finance, education and local government. So, a nurse, a bank manager, a bus driver and a teacher all work in the tertiary or service sector.

Triangular graph showing the occupational structures of selected countries

As you can see from the graph, there are big differences between the employment or occupational structures of different countries. Zaire has a very high percentage of people who rely upon primary industry. In Zaire, most people make a living from farming, either subsistence farming to produce their own food or producing cash crops for sale. There are also large numbers working in forestry and mining (for copper and cobalt).

In Russia, a high percentage work in manufacturing. In the UK and Germany, the tertiary (or service) sector is the main employer and few people work in the primary sector.

Whilst there are big differences between countries, you can see that the regional differences inside a country – the UK or Germany – are quite small. However, it is possible to see that some differences between regions do exist. You can see that some regions have fewer employees in the primary, secondary or tertiary sectors than others.

Note that Zaire and Indonesia can be classed as low income economies, Russia is middle income and Germany, Japan and the UK are high income.

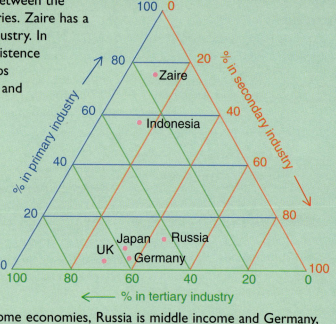

Most employees in the EU work in the tertiary sector such as financial services.

Employment statistics for UK, German and European Union regions

Region	Primary	Secondary	Tertiary
(Percentage of the work force employed in different sectors)			
Germany			
North Rhine-Westphalia (includes the Ruhr)	0.9	40.5	58.6
Berlin	0.2	41.0	58.8
Hamburg	0.2	25.6	74.2
UK			
North East England	1.8	40.6	57.6
South West England	3.5	30.0	66.5
Scotland	2.7	32.7	64.6
Other EU regions			
Brussels	0.0	19.0	81.0
Central Greece	28.3	30.3	41.4
North East Spain	3.5	41.8	54.7

STAGE 2: The old industrial regions

During the nineteenth and early twentieth centuries - from about 1820 to 1960 - parts of the UK and mainland Western Europe became industrial regions. The early years of this period are known as the **Industrial Revolution** when mass production in mills and factories began. These industrial regions usually had good supplies of power and raw materials. They concentrated on manufacturing for home and abroad.

The information which follows has maps showing the main industrial regions and includes details about the two regions on which we are focusing - North East England and the Ruhr in Germany.

On outline maps of North East England and the Ruhr, mark the main cities and add notes to show the main sources of power, raw materials and the location of the main industries.

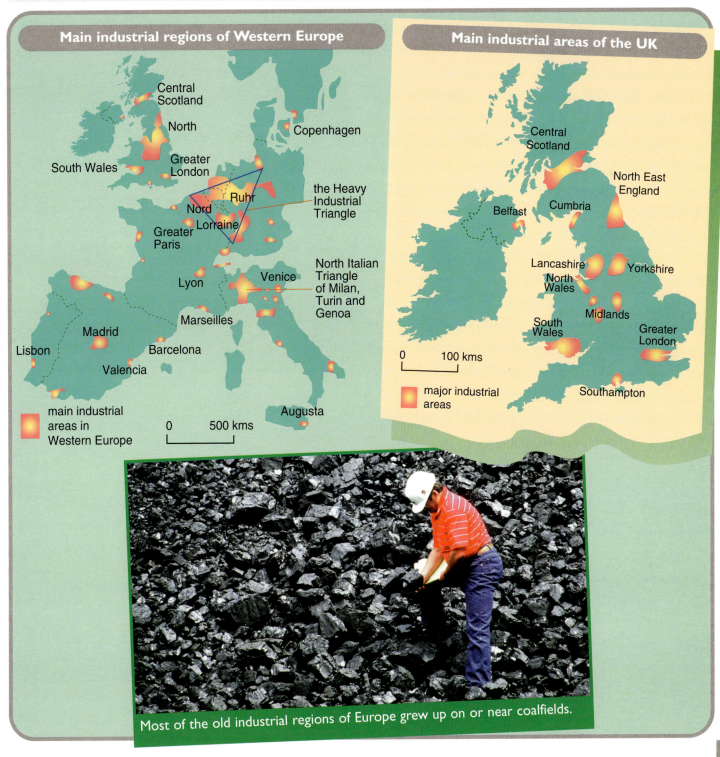

Most of the old industrial regions of Europe grew up on or near coalfields.

The Ruhr

The Ruhr is a tributary of the River Rhine which is Western Europe's most important waterway. Around the confluence of the Rhine and Ruhr a huge industrial conurbation has grown. It lies at the cross roads of Europe where the north-south River Rhine crosses ancient east-west routeways from France to Germany. The reason for the original industrial growth of this area was the Ruhr Coalfield. Today, over 10 million people live in this conurbation which includes the cities of Cologne, Dortmund, Dusseldorf and Essen.

The Ruhr is a classic example of a heavy industrial region based on coal. Today, it still produces most of Germany's coal, steel, petrochemicals and heavy engineering products.

The Ruhr coalfield produces very high quality coking coal, used for steel making and for steam power. After German unification in 1871 the region 'took off' economically. Coal production rose from 5 million tonnes per year in 1870 to 60 million tonnes in 1900. Canals were built to transport the bulky raw materials and heavy industry developed. The Rhine provided an easy transport link to Rotterdam and world markets beyond.

Alongside the coalfield there were deposits of iron ore which, together with limestone from the nearby hills of Sauerland, allowed the development of iron and steel making. The steel was then used for heavy engineering, including the major armaments manufacturer Krupps.

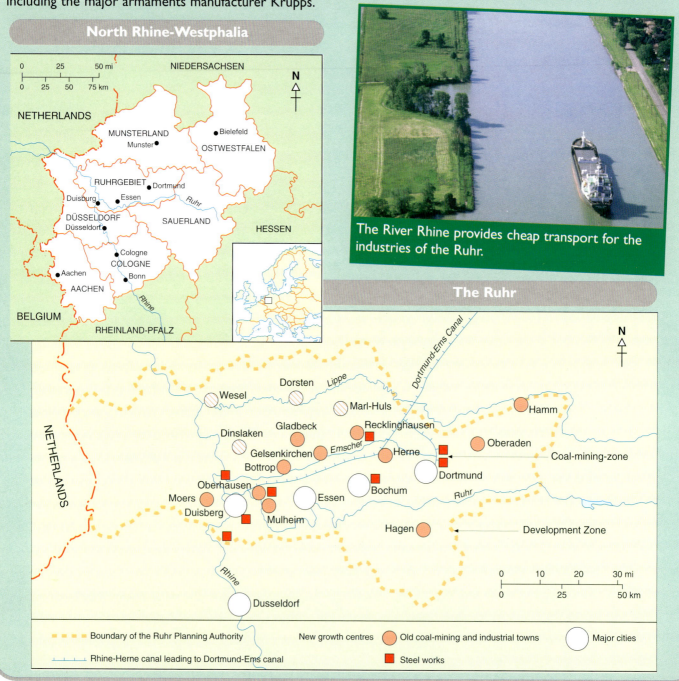

North Rhine-Westphalia

The River Rhine provides cheap transport for the industries of the Ruhr.

The Ruhr

- - - - Boundary of the Ruhr Planning Authority	New growth centres · Old coal-mining and industrial towns ○ Major cities
⊥⊥⊥ Rhine-Herne canal leading to Dortmund-Ems canal	■ Steel works

North East England

The major cities of North East England are Newcastle upon Tyne, Sunderland and Middlesbrough.

During the 19th century this region became important for its coal mining, steel making and engineering industries. Productive coalfields supplied steel works at Consett and Redcar. In turn, the steel was used to manufacture ships on the Tyne and Wear, railway locomotives at Darlington and engineering products over the whole region. A large proportion of all the world's shipping was built on the Tyne and Wear during the time that Britain dominated that industry. On Tyneside, the massive Armstrong factories became major producers of warships and guns. The firm amalgamated with Vickers shipyards to become Vickers-Armstrong, one of the world's greatest engineering companies. Today, Vickers still manufactures tanks on Tyneside.

The factors that helped the North East become one of Britain's leading centres for heavy industry were:

- an abundance of coal, close to the surface and easily mined in the Northumberland and Durham coalfield.
- supplies of iron ore and limestone from the Pennines and Cleveland Hills to supply the large steel works.
- river estuaries on the Tyne, Wear and Tees which were deep enough to handle the size of ships built in the 19th and early 20th century.

The North East of England was the world's first major centre of railway engineering.

Key:
- Railway lines
- Northumberland National Park
- Land over 200m
- Nuclear power station
- Alcan aluminium station
- Steelworks
- Major cities
- New town

0 5 10 mi
0 10 20 km

STAGE 3: Coal

Without coal the Industrial Revolution and the European dominance of world industry would not have happened. In the 19th century, Europe had the technology, the money and the power it needed to build its industries. The power came from coal. At first it was used to produce steam power and later to produce electricity.

Look at the following resources and then make some notes - perhaps in the form of a spider (or star) diagram - to describe and explain the growth and decline of the coal industry in North East England and the Ruhr.

The growth and decline of King Coal

During the Industrial Revolution in both Germany and the UK, the main source of power for the new industries was coal. It was burnt to produce steam which powered machinery, steam trains and steam ships. Later, massive amounts of coal were used in power stations to operate steam turbines and generate electricity. Whilst coal is still a very important fuel in both countries, it is no longer the main source of power. Transport now relies on oil, as do many other industries. Electricity is generated by burning gas and oil and also from nuclear power. Solar and wind power are also beginning to provide sources of renewable energy.

The disadvantage of coal is that it is bulky and therefore expensive to transport. Gas, oil and electricity are all cheaper to move. Coal is seen as a 'dirty' fuel because of the dust and soot it creates when burnt and also because of the carbon dioxide it releases. Other fossil fuels such as gas also release carbon dioxide, but not in the same quantities.

The decline of UK coalfields

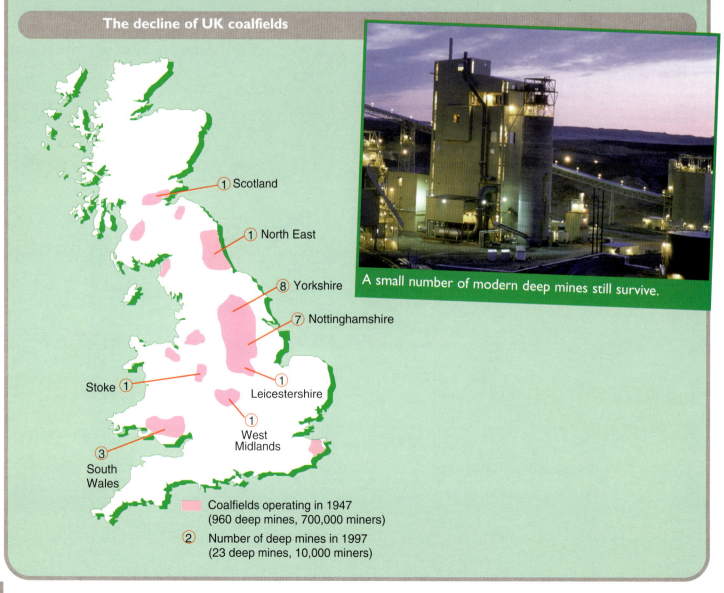

A small number of modern deep mines still survive.

Scotland ①

North East ①

Yorkshire ⑧

Nottinghamshire ⑦

Stoke ①

Leicestershire ①

West Midlands ①

South Wales ③

■ Coalfields operating in 1947 (960 deep mines, 700,000 miners)

② Number of deep mines in 1997 (23 deep mines, 10,000 miners)

The North East

Once there were hundreds of small pits in the North East of England. Coal from Northumberland and Durham was sent all over the world from the coal staithes (docks) on the Tyne and Wear. London and the South East relied on 'sea coal' from the North East, as did the great steel works and ship yards of the region.

But as the coal near the surface was worked out and cheap imports became available, the pits began to close. When the Conservative Government of the 1980's confronted the miners' unions, it accelerated the process. At the same time, other sources of cheaper energy such as North Sea gas and oil made many pits unprofitable.

Today, deep mining has largely been replaced by open cast mines. This controversial development requires few workers and produces much cheaper coal, but also creates huge scars on the landscape.

Cross section of the North East Coalfield

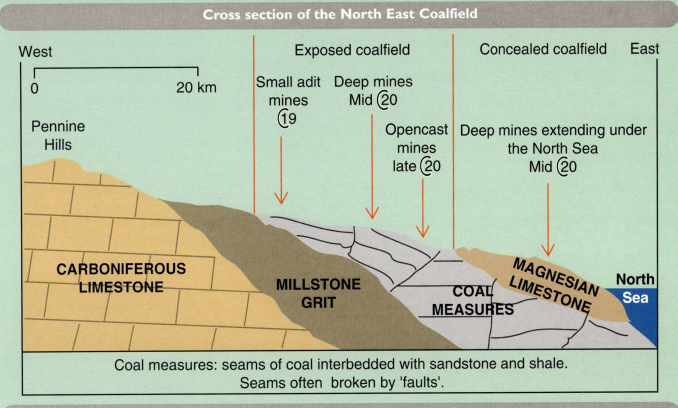

West   Exposed coalfield   Concealed coalfield   East

0   20 km

Pennine Hills

Small adit mines (19)

Deep mines Mid (20)

Opencast mines late (20)

Deep mines extending under the North Sea Mid (20)

CARBONIFEROUS LIMESTONE

MILLSTONE GRIT

COAL MEASURES

MAGNESIAN LIMESTONE

North Sea

Coal measures: seams of coal interbedded with sandstone and shale. Seams often broken by 'faults'.

Ellington: The last pit

In January 1995, the shock news reached the miners. With the loss of 2,000 jobs, the last pit in the North East was to close. These men thought that their jobs were safe. Following the 1984 miners' strike other pits had been closed but Ellington remained profitable. With coal seams extending out under the North Sea and enough reserves for years to come, the future looked secure. But in 1995 came the decision. British Coal declared the pit unprofitable and it was closed, like all the others in the North East. The great Ashington Colliery in 1988, Monkwearmouth in 1993 and now Ellington.

In Northumberland alone, 33 pits had closed since 1965 and 25,000 jobs had been lost. But some good news came later in the year. The pit was bought and reopened by RJB mining, a private company. RJB already owned the large pits in the Selby Coalfield, in Yorkshire, and the remaining pits in the East Midlands. Now, £13 million was to be invested at Ellington to open up new seams and supply the nearby Alcan works with 1 million tonnes of coal per year. The contract will last for at least five years with a promise of jobs for up to 400 miners.

RJB Mining operates a total of 20 deep pits (in addition to open cast mines); 85 per cent of its production going to electricity power generators. The company is confident that it has, at long last, halted the decline of the British coal industry. It seems unlikely that any new nuclear power stations will be built in the future (because of safety fears) and North Sea oil and gas production have peaked. Nevertheless, RJB knows that its coal must remain competitive with imports and that, during the next century, renewable sources of energy such as solar, wind and wave, will become important sources of power.

The Ruhr

Germany's major coalfields have always been in the Ruhr. Unlike in British coalfields, 85% of German coal now produced is **lignite** or brown coal. This is a less dense form of coal which generally produces lower heat and more CO_2 when burnt. But it is found in abundance and close to the surface where it can be mined easily and cheaply using open cast methods.

The Ruhr's reserves of **bituminous** (black) coal have become uneconomic to mine. It mostly comes from the deep (and therefore expensive to operate) pits in the north of the region. Where coal is still needed it can be bought more cheaply on world markets from countries as far away as Australia or China, or from the countries of Eastern Europe. In fact, German coal is four times more expensive than that on the world market.

The mines are only kept open because of large government subsidies. The German government thinks that this is worthwhile because the country has no oil or gas reserves and the use of nuclear power is politically unpopular. It does not want the remaining 100,000 coal miners to be unemployed. However, the government also realises that it cannot subsidise the industry for ever and wishes, over time, to reduce the level of support it gives.
The Ruhr Coalfield still contains an estimated 65 billion tonnes of coal reserves.

Cross section of the Ruhr Coalfield

North West   South East

Concealed coalfield   Exposed coalfield

Deep mines ⓶0

Shallow mines and adit mines ⓵9
Open cast mines ⓶0

River Lippe   River Ruhr

0   30 km

ALLUVIAL CLAYS

COAL MEASURES

CARBONIFEROUS LIMESTONE AND GRITS

SANDSTONE

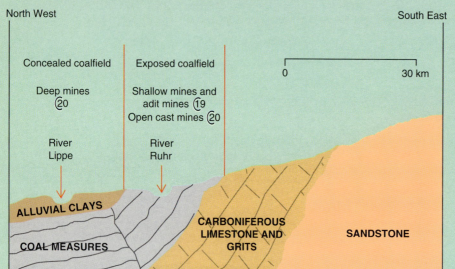

Open cast coal mining is much cheaper than deep mining.

Garzweiler 2

The Garzweiler open cast mine near Bonn produces 35 million tonnes of brown coal, or lignite, each year. The mine is owned by Rheinbraun, a subsidiary of Germany's largest electricity company, RWE. Brown coal supplies 85% of the energy for the Ruhr. But the supplies are running out and by 2006 the last brown coal at the site will have been extracted. As a result, the company has asked the government for permission to open a new open cast mine - Garzweiler 2 - next to the old one. The ruling Social Democrat local government granted permission but the Greens, a large party in Germany, are against it. Why?

- thousands of villagers will have to be resettled to make way for the mine
- brown coal contains high levels of CO_2 gases which pollute the atmosphere and add to the greenhouse effect. The German government has promised to cut CO_2 emissions by 30 per cent.

Rheinbraun argues that if it cannot operate Garzweiler 2, nearly 3,000 jobs will be lost in an area where unemployment is already high. They claim that new technology will reduce CO_2 (experts think this will be by about 10%). They also say that, since the Greens are also against nuclear energy, there is no alternative.

STAGE 4: Steel

If coal was the source of energy that fuelled the Industrial Revolution, then steel was the backbone of Europe's industry in the early twentieth century. The resources which follow show how the steel industry has changed and how this has affected North East England and the Ruhr.

Draw a diagram that shows the processes involved in steel making. Underneath, make notes that explain why coastal locations have become the most efficient places to produce steel. Briefly describe how steel making has changed in North East England and the Ruhr. Explain why so many older steel works have closed.

Steel Making

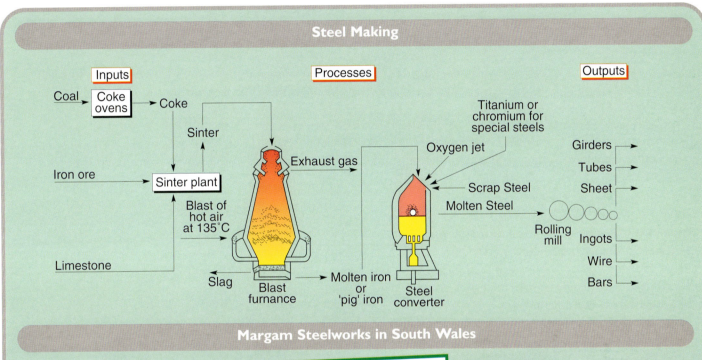

Inputs — Processes — Outputs

Coal → Coke ovens → Coke
Sinter
Iron ore → Sinter plant
Blast of hot air at 135°C
Limestone
Slag
Blast furnace → Molten iron or 'pig' iron → Steel converter
Exhaust gas
Titanium or chromium for special steels
Oxygen jet
Scrap Steel
Molten Steel → Rolling mill

Outputs: Girders, Tubes, Sheet, Ingots, Wire, Bars

Margam Steelworks in South Wales

This is an example of an **integrated steel works**. All the processes are carried out on one site in a continuous flow production. This saves on transport and fuel costs compared with older steel mills where iron and steel making were not integrated, ie they were separate processes.

The plant is built on the coast next to Port Talbot docks so that imported iron ore can be unloaded from bulk carriers. This is much cheaper than unloading the ore onto trains and then transporting it inland. By using imported ore, the steel producers can 'shop around' the world and obtain the lowest cost ore - rather than being tied to one local producer. The steelworks uses coal from the nearby South Wales Coalfield and also imports coal.

The plant is built on a large flat area of land because the processes are so large scale. At one end of the site, near the docks, the imported ore is moved by conveyor belt into blast furnaces. The molten 'pig' iron is then converted into steel. This is then rolled into sheets or girders, depending on customer requirements.

Steel in North East England

In 1875, Britain produced a very large proportion of the world's iron, steel and engineering goods. British ships, locomotives, rails, bridges and machinery were exported all over the world. Steel was needed not just in Britain but throughout the British Empire - in Canada, India, Australia and Africa.

In the North East, local coal and raw materials encouraged the growth of a large steel industry. There was plenty of local labour and a large market in the shipbuilding and engineering industries of Tyneside, Wearside and Teesside. A number of iron and steel works were built, with the largest at Consett in County Durham. These works were located on the coalfield near to iron mines. In the late nineteenth century large deposits of higher grade (ie, more pure) iron ore were discovered in the Cleveland Hills, just south of the River Tees. Large steel works were then built on Teesside and the coal was brought the short distance from the coalfield by rail. The old inland steel works were gradually closed down with only Consett remaining. They were inefficient and could not compete with modern steel mills, particularly in Taiwan, South Korea and Japan.

During the second half of the 20th century new supplies of high grade iron ore started to be imported. This is transported by giant ore carriers which dock on Teesside. A modern integrated steel works was built at Redcar. Integrated steelworks, on the coast, are more efficient than the older mills like Consett. In the 1980s Consett was finally closed with a loss of 4,000 jobs.

Today, Redcar is one of only four large scale integrated steel works in Britain. In the 1990s these steel works, under the management of British Steel, have become very efficient. The labour force has been cut, new investment has taken place and the company is one of the most efficient steel producers in the world - once again exporting to many countries.

Consett Steelworks

The old steelworks were closed down because they were too small to be efficient. Modern integrated steelmaking was instead transferred to the coast at Redcar where imported ores could be brought by bulk carriers.

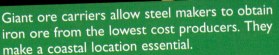

Giant ore carriers allow steel makers to obtain iron ore from the lowest cost producers. They make a coastal location essential.

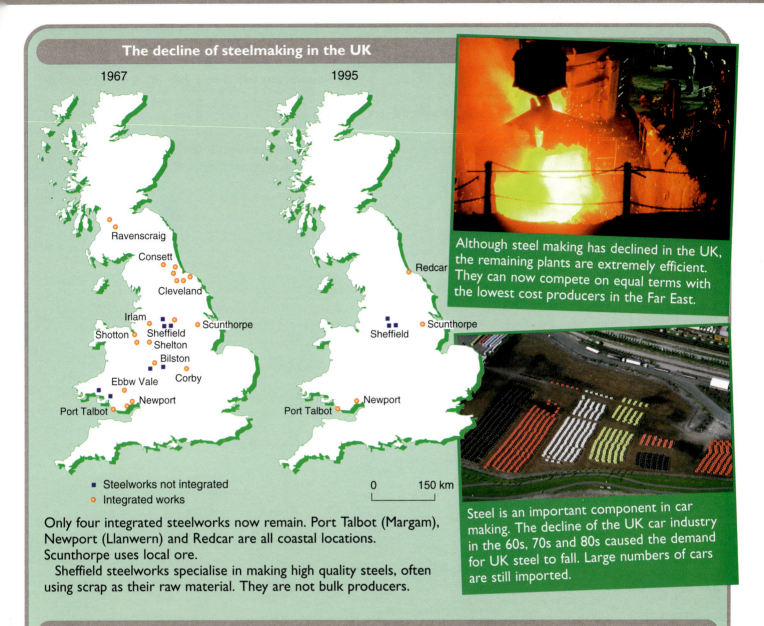

The decline of steelmaking in the UK

1967

Ravenscraig
Consett
Cleveland
Irlam
Shotton Sheffield Scunthorpe
Shelton
Bilston
Ebbw Vale Corby
Newport
Port Talbot

1995

Redcar
Scunthorpe
Sheffield
Newport
Port Talbot

■ Steelworks not integrated
○ Integrated works

0   150 km

Although steel making has declined in the UK, the remaining plants are extremely efficient. They can now compete on equal terms with the lowest cost producers in the Far East.

Steel is an important component in car making. The decline of the UK car industry in the 60s, 70s and 80s caused the demand for UK steel to fall. Large numbers of cars are still imported.

Only four integrated steelworks now remain. Port Talbot (Margam), Newport (Llanwern) and Redcar are all coastal locations. Scunthorpe uses local ore.

Sheffield steelworks specialise in making high quality steels, often using scrap as their raw material. They are not bulk producers.

Steel from the Ruhr

20% of all the steel produced in the European Union is made in the Ruhr. Heavy industry boomed in this area during the late 19th century. There were a number of reasons:
- there were large local deposits of coal and iron ore
- there was cheap water transport on the Rhine
- there were good rail, canal and road links
- there was a large market in the newly unified country of Germany
- the Ruhr has a central location in the industrialised and heavily populated markets of NW Europe
- the growing army and navy of Germany demanded increased steel production in the period up to 1914.

Today, the Ruhr is no longer an ideal location for steel production. The local coal is expensive compared with imports from Poland or the USA. Iron ore has to be imported because the local area no longer has high grade reserves. Some comes from the Lorraine area of Eastern France and some comes by barge along the Rhine from Rotterdam where it is brought in ore carriers from overseas. German labour is expensive, especially compared with Taiwan, South Korea or Brazil - all major steel producers.

But there is a large local market and so the industry remains. This is an example of **industrial inertia**. It means that once an industry is established in an area, with a lot of money invested in plant and many jobs dependent upon it, it will often remain in that location even when it is no longer cost effective.

Some old steel works have closed and large new integrated steel works have been built on the banks of the Rhine at Duisberg. Smaller, specialised steel plants are located at Bochum and Dortmund. The German government gives these works subsidies so they can buy local coal.

STAGE 5: Industry in decline

So far we have seen that heavy industry in the North East and the Ruhr was based on an abundant local supply of coal for power, together with the strength of the steel industry.

In the information that follows, we consider why these regions have lost so many industries and jobs since the 1960s.

Use the data provided to draw graphs showing employment in the main industries of the North East. Make a list that summarises the reasons for the decline of the major industries, eg coal, steel, engineering and shipbuilding in both the North East and the Ruhr.

'Toon army on the dole' - job losses in the North East

In a region where 150,000 men used to work for British Coal, now there are none. When the last deep pit closed at Ellington in 1995 it seemed like the end of an era - even though the pit later reopened in private hands. Not only have there been massive job losses in coal mining, but also in steel making. During the 1980s the world famous Consett steel works closed with the loss of thousands of jobs. Now there is no sign of the old steel works and the town is more famous for making snack foods.

Why all the closures? Coal became too expensive to mine; imported coal was cheaper and there was conflict between the miners' union - the NUM - and the government in the 1984 miners' strike. The government at the time decided that it did not wish to subsidise or support the mining industry.

Today there are fewer uses for coal. Most people have gas central heating and there are no steam trains or steamships using coal fired boilers. The decline of the shipbuilding and engineering industries reduced the market for steel. Power stations generating electricity increasingly use oil or gas, and nuclear power has been given a guaranteed slice of the energy market. And whilst once Britain's shipyards supplied the world, now the industry has all but gone. The world buys its ships from lower cost producers in South Korea and Taiwan. Most of the British rivers famous for shipbuilding, the Clyde, Mersey, Tyne and Wear, are too shallow and narrow for building the modern super tankers.

Employment in the North East			
Year	Mining	Chemicals	Shipbuilding
1947	150,000	50,000	67,000
1959	130,000	58,000	64,000
1970	49,000	60,000	39,000
1983	29,000	48,000	20,000
1990	13,000	35,000	4,000
1996	500	30,000	4,000

Most of the world's supertankers are now built in the Far East, in modern shipyards with lower labour costs.

The decline of shipbuilding on the Tyne has caused many job losses - not just in the shipyards but also in those industries that used to supply raw materials and components.

From cradle to grave - death of the Ruhr

The Ruhr, once the cradle of Germany's huge arms industry, is in terminal decline. A catastrophe is looming according to Werner Nass, a union official at the giant Krupp - Hoesch steel plant in Dortmund.

Today, Dortmund has all the excitement and glamour of a wet Sunday afternoon. The mood in the pubs and labour exchange is gloomy. Unemployment rates are the highest in Germany. Bend Kowalski sips his 'Dortmund' Beer which is no longer brewed in the city since a new modern brewery was built further north. Herr Kowalski is 56 and has been retired early by the Krupp - Hoesch steel plant which has a policy of retiring workers at 55 to reduce the work force.

Labour costs are higher here than elsewhere in Europe and it is difficult to attract investment. As the old steel and chemical plants become out of date they are closed. German coal is too expensive and iron ore has to be imported. Only government subsidies keep the coal mines and steel works open. But for how long?

Since the 1960s, 500,000 jobs have been lost in mining and another 300,000 in steel, chemicals and heavy engineering. The decline goes on.

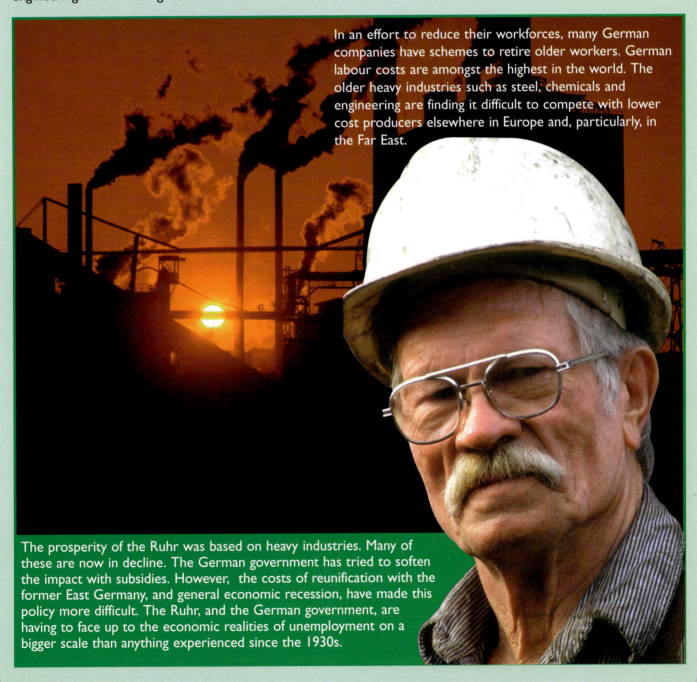

In an effort to reduce their workforces, many German companies have schemes to retire older workers. German labour costs are amongst the highest in the world. The older heavy industries such as steel, chemicals and engineering are finding it difficult to compete with lower cost producers elsewhere in Europe and, particularly, in the Far East.

The prosperity of the Ruhr was based on heavy industries. Many of these are now in decline. The German government has tried to soften the impact with subsidies. However, the costs of reunification with the former East Germany, and general economic recession, have made this policy more difficult. The Ruhr, and the German government, are having to face up to the economic realities of unemployment on a bigger scale than anything experienced since the 1930s.

STAGE 6: Counting the costs

When major industrial areas go into decline, there are massive **social**, **economic** and **environmental costs** to bear. In both the North East of England and the Ruhr, the decline of industries and the job losses that follow have caused severe problems since the 1960s.

Using the information that follows, draw a spider (or star) diagram that illustrates the main problems that arise when a region's industries decline.

Social costs

- High rates of unemployment, especially amongst skilled men who worked in heavy industry. Often women find it easier to obtain employment in lighter manufacturing and service industries - so creating domestic problems as men remain on the dole.
- Fewer job prospects for less qualified school leavers.
- Social unrest. Parts of Newcastle and North Tyneside have suffered from rioting by groups in depressed areas. Some of the rioting was racist with Asian families and businesses targeted by the criminals. In the Ruhr, right wing extremists have attacked Turkish migrant workers and the 'Ossis' - East Germans who moved into the area looking for work since reunification in 1990.
- Crime. There is some evidence that high rates of unemployment - particularly long term - increase the crime rate.
- **Depopulation**. People leave the area looking for work elsewhere. Families are split and there is an increasing proportion of older people left behind. When the population falls the whole regional economy suffers. There is less demand for other goods and services and the tax revenue of local government falls.

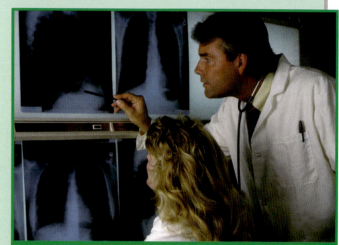

People living in areas high unemployment and declining industries tend, on average, to be less healthy than those in better off areas.

- Health. There is a clear connection between unemployment and poor health. Unemployed people suffer from more medical problems than those in work.

Many jobs in the service industries have traditionally been carried out by women. Men who have spent a lifetime working in heavy industry find it difficult to adjust to such jobs. As a result, in an increasing number of families the 'breadwinner' is a woman. Unfortunately, many jobs in the service sector are low paid or part time, so families find their incomes are low.

When miners become unemployed they often find it difficult to get new work. Their skills and expertise gained in mining is not often needed in new industries.

Economic costs

- Unemployment costs money. Social security benefits are high in both areas.
- Government revenue from taxes fall as industries close and people lose their jobs.
- The run down and depressed image of the regions make it difficult to attract new industries. When one industry declines it causes **knock on effects** - suppliers and component manufacturers also suffer, and other local businesses are affected because unemployed people have less money to spend. The whole regional economy can go into a downward spiral. Government grants and other incentives have to be offered to attract new industries.
- House prices fall in the industrial areas, some houses become impossible to sell - so 'trapping' the unemployed in the area.
- Government subsidies to keep industries alive cost large amounts of taxpayers money. In Germany, every job in a pit costs the government DM100,000 per year.

Job losses in one industry can have a knock on effect on others. It is very difficult for regions with depressed economies to break out of the downward spiral.

Environmental costs

- Derelict factories are dangerous and unsightly.
- Waste land is created, often polluted with industrial effluent and toxins.
- Old mining areas, waste tips and steel plants have to be reclaimed. Sometimes they are used for creating country parks or land for housing.
- Areas of older housing fall into disrepair because people cannot afford to renovate and the council has little income from local taxes to help them.

Clean up costs can be very expensive. When factories close down it is often the local community which is left with the expense of cleaning up the old sites.

Toxic waste from derelict factories can poison water supplies.

STAGE 7: Regeneration

When an industrial area declines it causes problems for the people and the government. Poverty, unemployment and crime increase, people move away - and they often blame the government.

In the resources which follow, we look at how the UK and German governments, and the European Commission, have tried to help attract new businesses to the regions where jobs have been lost.

Using information provided in this and previous Stages, make two lists - one to show the sorts of industries a school leaver in the North East or the Ruhr could have worked in during the 1960s and an equivalent list for the 1990s. How are they different?

Briefly summarise the reasons why the future might be better for the two regions.

The New North East

This was once a region with heavy industry in decline and massive job losses in coal, steel and shipbuilding. But in the 1990s the North East has seen a revival. There are still problems with high rates of unemployment in areas of Tyneside, in particular, and a massive gap in living standards between the rich and poor, but new jobs are being created.

In 1995 and 1996, there were a number of optimistic developments for the North East with the announcement of more investment by the Japanese firms Nissan and Fujitsu and the giant German electronics group Siemens.

So, how has the North East managed its success and what are the signs of regeneration in this area?

Attractions:

Government grants to new firms (called **incentives**) from both the UK and the EU. The region receives help because of its status as a UK Development Area (due to its high unemployment) and because it is eligible for European Coal and Steel Community funding. This latter fund is specially designed to help those regions which have suffered from the decline in coal and steel.

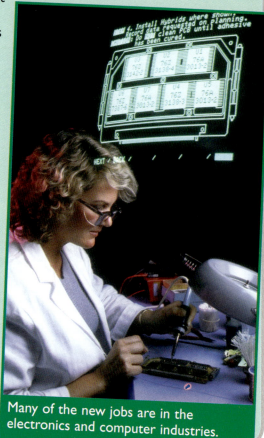

Many of the new jobs are in the electronics and computer industries.

- Local governments in the area provide free advice and help with planning applications for new businesses.
- Low production costs with lower wage rates than in Japan or Germany.
- A plentiful supply of skilled and semi skilled workers, including engineers.
- A large market - Britain is the sixth largest consumer of electronics goods in the world and the third largest in Europe. The Single European Market means that firms have free access to all the EU economies.
- The availability of cheap land for building - some of it is reclaimed derelict land close to areas of population.
- Container and ferry terminals on the Tyne and Tees give direct and fast sea transport to mainland Europe.

New jobs in the North East

- Procter and Gamble. US based pharmaceutical and chemical company. Manufactures detergents. UK headquarters at Gosforth in Newcastle.
- Nissan. Japanese car manufacturer. Largest European plant in Washington, County Durham.
- AMEC Offshore. American owned manufacturer of offshore oil rigs at Wallsend on Tyne.
- Metro Centre, Gateshead. The UK's largest out of town shopping mall. A major employer and generator of income.
- Komatsu. Korean manufacturer of excavators at Birtley, Gateshead.
- Fujitsu. Japanese domestic appliances and electrical goods, Newton Aycliffe.
- Newcastle Airport. Rapid expansion since 1980 into a major UK airport helping to attract industry and employing 500 people.
- DSS. An example of the decentralisation of government offices. The new offices at Longbenton employ 2,000 people.

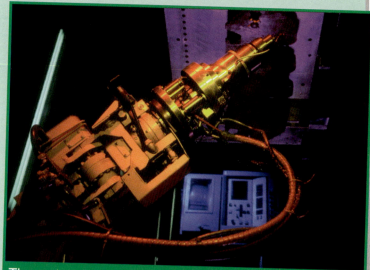

The engineering companies that have survived use high tech production methods. Investment in robots might mean that fewer workers are needed, but production is efficient and error free.

The top manufacturers in the North East

	Employees	Industry
Rolls-Royce Power	19,000	Electricity generators
Greggs	12,000	Baking
Vaux	9,000	Brewing
ICI	8,000	Chemicals
Claremont Garments	6,000	Clothing
Procter and Gamble	5,000	Detergents
Nissan	4,000	Cars
Sage	1,800	Software
Amec	1,700	Oil rigs
Redpath	1,300	Engineering
Interconnections	1,300	Circuit boards
Perstop - Ferguson	1,200	Laminates
Fujitsu	900	Computers and electronics
Barbour	850	Clothing
Caterpillar	800	Trucks
Flymo	750	Engines
Hydro Polymers	500	Resins
Alcan	475	Aluminium
Komatsu	420	Excavators
Also		
Vickers	Not available	Tanks
Samsung	3,000 jobs planned	Televisions

Revitalising the Ruhr

Some might think that the economy of the Ruhr is finished - but the regional government of North Rhine-Westphalia has other ideas. As 800,000 jobs have been lost in coal and steel, so the government has helped create 800,000 jobs through investment in service industries, conservation and recycling industries.

With a population of 20 million, North Rhine-Westphalia cannot afford to let its economy disintegrate. As old industries decline, so new ones are attracted. Foreign investment by companies such as Hitachi from Japan has helped, but the government has also directly intervened:

- Higher Education. In 1960 there were 63,000 students. Today there are over 500,000. A highly educated work force helps attract new industry and education itself creates jobs.
- Recycled products. North Rhine-Westphalia is a world leader in recycled paper, glass and other products. The industry has a value of DM20 billion per year.
- Telecommunications. Veba, Germany's largest industrial group manufactures mobile phones and other telecommunications equipment in the area. 26% of all German TV programmes are made in the region, mostly in Cologne and Dusseldorf.
- Media. The region is the centre for the German radio, TV and newspaper industries. Together they are the third largest employer.

Unlike the North East of England, the Ruhr still retains much of its old industrial base. It still produces 60% of German coal and has large reserves. Steel, chemicals and textiles are still important employers.

Recycled aluminium. 'Green' recycling policies have led to a massive new industry.

Hitachi comes to the Ruhr

When Hitachi Seiki, one of Japan's leading makers of machine tools, decided to build a factory in Europe it took a long time to decide on its site. Eventually the town of Krefeld, in the Ruhr, was chosen.

Krefeld is close to Dusseldorf where over 5,000 Japanese now live - brought here by the growth of Japanese companies in the area. More importantly, Krefeld is within easy reach of the 20 suppliers of components which Hitachi needs - so reducing transport costs and ensuring supplies. By contrast, when Hitachi was considering its location, only 3 of the 20 component suppliers could be found in the UK.

While labour costs in Germany are high, Mr Hirario - the managing director of Hitachi Deutschland - claims that high productivity and good labour relations compensate for this. And the Japanese have a high regard for German engineering skills.

The Ruhr has become an important centre for Japanese electronics companies.

Perhaps of prime importance is the location of Krefeld at the heart of the huge European market. Other firms to come to the Ruhr include Mitsubishi Semiconductor which makes computer components. Unfortunately, many German companies are leaving their old premises in the Ruhr to establish new factories in Eastern Europe, eg in Poland or the Czech Republic where costs are much lower.

Industry in North Rhine-Westphalia, 1995

Breakdown in rank order of turnover

Sector	Turnover	Employees
Chemicals	DM70bn	170,000
Machine building	DM55bn	241,000
Food industry	DM48bn	112,000
Electronics	DM40bn	175,000
Vehicle construction	DM37bn	108,000
Building and materials	DM27bn	118,000
Railways	DM26bn	85,000
Petroleum	DM23bn	6,000

10 largest enterprises, 1995

Veba AG, Essen	manufacturing
RWE, Essen	energy
Telekom, Bonn	telecommunications
Bayer, Leverkusen	chemicals
Thyssen AG, Duisberg	steel and machinery
Metro-Gruppe, Dusseldorf	trading
Rewe-Gruppe, Cologne	retailing
Aldi Gruppe, Mulheim	trading
Ruhrkohle AG, Essen	coal mining
Mennesmann, AG, Dusseldorf	machine, automotive, electronics

The top industries in North Rhine Westphalia 1960

Textiles	1
Coal mining	2
Food processing	3
Steel	4
Chemicals	5
Construction	6
Machinery	7
Electronics	8

So far, the Ruhr has retained its older industries - such as chemicals - better than the North East of England.

STAGE 8: Review

By now, you will have a collection of notes, maps, graphs and diagrams on the changing employment and industrial pattern of the North East and the Ruhr. You are asked to organise all your information and then write an essay under the title: 'The changing Industry of North East England (or the Ruhr in Germany)'.

Before you start writing, make a plan of what your essay will include. You might start with a spider / star diagram or flow chart.

Glossary

All the terms listed below are explained in this Enquiry. In each case, write a definition of the term and, if possible, give at least one example.

Heavy industry	**Manufactured goods**	**Depopulation**
Occupational structure	**Industrial Revolution**	**Economic costs**
Primary sector	**Lignite**	**Environmental costs**
Secondary sector	**Bituminous coal**	**Knock-on effects**
Tertiary sector	**Integrated steel works**	**Regeneration**
Service industry	**Industrial inertia**	**Government incentives**
Raw materials	**Social costs**	

to the student

Many factors which once limited where a factory could be sited have now changed. Perhaps the biggest shift has been in transport and communications which are now much faster, cheaper and more efficient than in past decades. These changes have encouraged companies to produce and market their products on a global scale - a process sometimes known as *globalisation*. Political changes such as the lowering of import controls and other barriers to trade have also helped this trend.

While older, heavy industries such as iron and steel have declined, newer *light industries* such as electronics and consumer goods have grown.

In this Enquiry we look at new types of industry, their characteristics and the factors which affect their location - both in the UK and in other parts of the world.

questions to consider

1 What factors affect where a factory is located?
2 How have these factors changed over the past decades?
3 What are the main features of the 'new' industries?

key ideas

Most new industries are *footloose*. This means that they are not limited by having to be close to their source of raw materials or fuel. They can be situated almost anywhere. Often, new industries are located near their markets. In other words, near the centres of population where consumers live.

Few new industries employ the huge numbers of workers that the old steel, chemical and engineering industries once did. Instead, they are *automated*. In other words they use computers and robots to do much of the work previously done by hand.

Many of the new industries are on *greenfield sites* outside urban areas but close to motorways. Here they have room to expand and are not affected by urban congestion.

activities

Using information from the facing page:

Working with a partner, construct a table that shows the main differences between the old and new industries. Draw up your table with the following headings: Location, Workforce, Raw materials, Transport and Site.

THE NEW INDUSTRIES FOOTLOOSE AND MOBILE

Large, old factories surrounded by housing (Inner City)

'Heavy Industry' making bulky goods eg ships, steel

Need large amounts of bulky raw materials eg coal, ore, oil

Employ large numbers of men

← FROM THIS

Cause noise, air, traffic pollution

Close to railways, canals or ports for cheap transport of bulky goods

Often found on or near coalfields

Produce goods which are small, light, and with a high value

Modern, single storey buildings built from 1970's onwards

Main source of power is electricity, available anywhere and so 'footloose'

Many service industries including printing, research, banking

Situated close to their market or with good access to motorway or airport

TO THIS →

Often located with other factories on purpose built 'industrial estates' on the edge of towns

Employ fewer men and more women

Mainly use road transport for delivery of raw materials and distribution of goods close to main roads

enquiry

Industrial location

This Enquiry provides you with examples of new industries and explains the factors behind their location. Some are in the UK but others are on a world scale. Goods made by the biggest manufacturing companies such as Ford, IBM, Sony and Coca-Cola are produced in every continent. They are produced, advertised and sold on a global scale. For example, a Ford car sold in the UK might be made in Germany, Spain, Belgium or the USA and it will contain parts made in Japan and other countries.

The outcome of this Enquiry will be a decision making exercise in which you will be asked to recommend, with reasons, the site for a new factory. As you work through the Stages of the Enquiry, you will build up your knowledge and understanding of the issues involved.

STAGE 1:   What new industries?

The resources which follow contain information on some of the new and existing industries of the 1990s. Look at the new developments outlined below and make a brief note that describes each one. To what extent are the developments examples of globalisation, ie the trend towards global marketing and production?

Copy the graph showing the world production of televisions onto a large piece of graph paper. Then use the data in the table to mark on the output of the other countries that are named. Mark on a map where the leading producers are located. Briefly describe the main trends shown in the data.

Using the case study of the BOC Group, make some brief notes on why the company chose to locate a new factory in south east China.

Nissan to invest another £70 million in Sunderland

Nissan, the Japanese car maker, is to invest another £70 million in its Washington car plant, near Sunderland in the North East of England. This will enable the company to manufacture the estate car version of its Primera car instead of importing it from Japan, This will create another 150 jobs at the plant.

Hyundai chooses Fife

Hyundai Electronics from South Korea is to build two new factories at Dunfermline in Scotland. These factories will eventually provide 2,000 new jobs making semiconductors for the European **electronics** market. Semiconductors are used in most consumer goods including 'brown goods' (eg TVs, hi-fi, computers) and 'white goods' (eg microwaves, washing machines) and motor vehicles.

New orders for British Aerospace, 2,000 jobs secure

The British flagship in the **aerospace** industry has won a multi-billion pound contract to supply Malaysia with fighter aircraft and an air defence system. Thousands of jobs at British Aerospace plants in Bristol and Warton, near Preston, will now be secure for several years.

Photronics comes to Trafford Park, Manchester

US company Photronics chose Manchester for its new factory to be close to the motorway system and the international airport. The company manufactures parts for the semiconductor industry. Semiconductor wafers are used in most new electronic goods from mobile phones to aeroplanes. Photronics other factories are in California, Texas and Connecticut. 270 jobs will be created in Manchester.

Sandoz invests in Ireland

Swiss **pharmaceutical** company Sandoz is to invest in a new plant at Ringaskiddy, County Cork, bringing hundreds of new jobs. It will produce medicines and medical drugs for the whole Western European market.

| Television production (thousands) | | | | | 1992 | |
	1960	1970	1980	1992	% 100	Rank Order
World	25,000	44,000	71,000	128,000		
Azerbaijan	-	-	-	5,913	4.6	5
Brazil	183	726	3,254	3,500	2.7	9
China	0	0	2,492	28,678	22.4	1
Japan	3,578	12,488	15,205	12,000	9.4	4
South Korea	0	114	6,819	16,311	12.8	2
Malaysia	0	44	157	5,553	4.3	6
Russia	-	-	-	4,500	3.5	7
Ukraine	-	-	-	3,600	2.8	8
United Kingdom	2,141	2,214	2,364	3,400	2.7	10
USA	5,611	8,298	10,320	13,352	10.6	3

Note: Azerbaijan, Russia and the Ukraine were part of the former Soviet Union. Although production figures are not available for these new countries before 1992, the Soviet Union as a whole had been a major producer.

World output of TVs

In 1960 the UK was the third largest producer of televisions in the world. By 1992 it had fallen to tenth place. However, its overall production had increased, thanks mainly to Japanese and other Far Eastern manufacturers setting up factories here.

It is relatively cheap to transport the components of televisions, but the finished sets are bulky and fragile - making them expensive to transport. Therefore it makes economic sense to produce near the market.

The bigger the market and the richer the consumers, the more likely that producers will set up their factories.

Labour costs are also an important factor in deciding where to set up production. The manufacture of televisions is labour intensive - in other words it takes a lot of labour to make each set. The low wages paid to Chinese workers have been influential in locating TV manufacturing plants in that country.

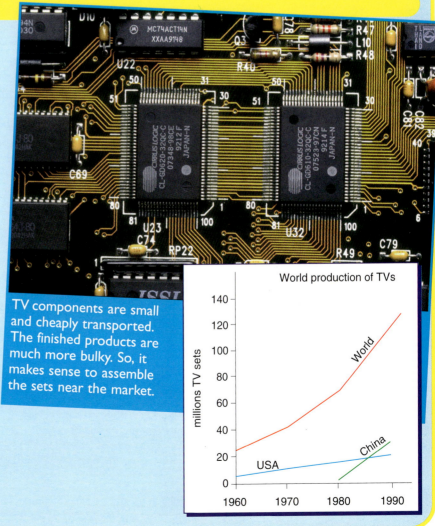

TV components are small and cheaply transported. The finished products are much more bulky. So, it makes sense to assemble the sets near the market.

World production of TVs

BOC Group

BOC is a **multinational** chemicals company based in the UK. It manufactures in over 60 countries and employs 40,000 people world wide. Sometimes multinational companies, ie those producing in more than one country, are called **transnational** corporations (TNCs).

The main part of its business is producing gases which are used in industrial processes and in health care. These gases include oxygen, nitrogen and carbon dioxide. They are used in the manufacture of just about everything - from microchips, steel, plastics and food to light bulbs. In health care, BOC gases are used in four out of every ten anaesthesia machines in use world wide.

Each year the company spends almost £100 million on **Research and Development** (R&D). This involves inventing, discovering and applying new materials and processes. It is vital for the company to spend this money, otherwise it will fall behind its competitors.

Because industrial gases are bulky and expensive to transport in cylinders, BOC sets up its manufacturing plants near to its customers. Therefore it is currently expanding in markets which are growing fastest. One such is in south east China, across the border from Hong Kong. Here, in the Guangdong Industrial Zone, massive investment is taking place by multinational companies eager to have a foothold in the 1 billion Chinese market. BOC's new plant will supply the specialist gases that these new industries require.

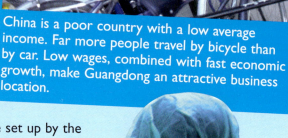

China is a poor country with a low average income. Far more people travel by bicycle than by car. Low wages, combined with fast economic growth, make Guangdong an attractive business location.

The choice of the Guangdong site was made for several reasons. The East Asian economies are growing at a rate of almost 8 per cent per year. Their demand for industrial gases is rising at twice that rate. The special economic zone set up by the Chinese government allows companies to set up with little red tape or restrictions. Wages are low, even by Asian standards and the government gives tax concessions for the first few years of operation. The local officials were helpful in finding a site that was central and close to the road network serving the whole industrial region.

Research and development is vital for modern multinational companies.

STAGE 2: Motor World

The world's economy is based, to a very large extent, on the motor car. Car production and ownership continue to rise dramatically. Some of the world's biggest corporations either manufacture motor vehicles (eg, General Motors which owns Vauxhall in the UK) or the fuel to run them (eg, Exxon, or Esso in the UK).

The resources which follow show where some of the major vehicle manufacturers are based. One particular example of Japanese investment in the UK car industry is described. (**Investment** includes expenditure on plant, machinery and training.)

Make a note on where the motor vehicle industry is growing most rapidly. Where is it likely to expand in the future? Explain your reasons.

Then, using the diagram of 'What it takes to build a car', explain how a new car factory might affect, and be affected by, the location of other factories.

Complete the tasks outlined above the Ordnance Survey map extract.

Trends in world car production (thousands)

	1948	1970	1980	1993	1993	
					% 100	Rank Order
World	4,640	22,550	28,999	36,000	100	
Canada	167	923	847	1,100	3	9
France	100	2,458	3,487	2,921	8	4
Germany	33	3,655	3,688	3,926	11	3
Italy	44	1,720	1,445	1,130	3	8
Japan	0.4	3,179	7,038	8,682	24	1
South Korea	0	13	58	1,528	4	5
Russia	20	344	1,327	1,000	3	10
Spain	0.2	455	1,048	1,506	4	6
UK	335	1,641	924	1,375	4	7
USA	3,909	8,505	6,376	5,700	16	2

Note: Europe combined manufactures about 30% of world cars. Figures for Russia pre 1993 refer to the USSR.

How British is a Rover car?

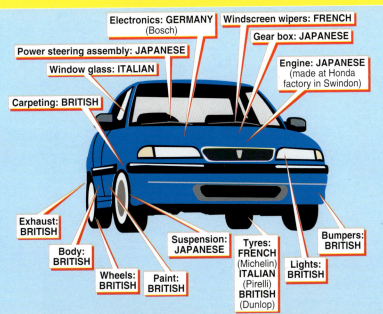

Electronics: GERMANY (Bosch)
Windscreen wipers: FRENCH
Gear box: JAPANESE
Power steering assembly: JAPANESE
Engine: JAPANESE (made at Honda factory in Swindon)
Window glass: ITALIAN
Carpeting: BRITISH
Exhaust: BRITISH
Body: BRITISH
Wheels: BRITISH
Paint: BRITISH
Suspension: JAPANESE
Tyres: FRENCH (Michelin) ITALIAN (Pirelli) BRITISH (Dunlop)
Lights: BRITISH
Bumpers: BRITISH

The Rover company is now owned by the German BMW company. However, the cars are still assembled in British car plants in the Midlands.

The world car

Making and running motor vehicles is one of the biggest industries in the world. In the USA alone, 14 million people work in the motor industry and its service industries, eg filling stations, garage repairs and sales. The industry accounts for 15% of all US business. On a global scale the car industry consumes 20% of world steel and 50% of world rubber production. It also has a major impact on the development of metals, plastics, paints, oil refining and electronics.

This giant industry demands massive investment and huge sales to keep it going. But in North America, Japan and Europe the market is saturated. Whilst there is a huge demand there is too much capacity - the factories can make 30% more cars than are needed. So where does the industry expand?

A look at the world map showing Vehicle Density tells you where there might be potential new markets. Why is vehicle ownership relatively low in Africa, Asia and South America? The maps showing world population density and growth rates (pages 7 and 9) tell you how enormous the potential market is. What must happen in countries like India and Brazil before vehicle ownership increases to North American levels? And what will be the consequences if it does?

For companies like Ford, which has invested over $6 billion in developing the Mondeo, a 'world car', it is vital that it opens up these new markets. But what effect will this have on world resources and air pollution?

Car manufacturing is one of the world's major industries. The big producers are trying hard to open up new markets.

World vehicle density, 1995 (figures are cars per 1,000 inhabitants)

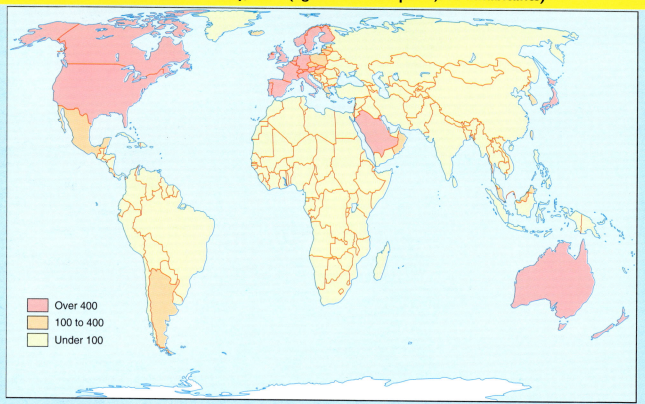

Over 400
100 to 400
Under 100

Japanese investment in Europe

Nissan
Sunderland, UK
Capacity: 300,000 cars a year.
Output 205,000 in 1994.
£1bn investment

Honda
Swindon, UK
Capacity: 150,000 cars a year.

Isuzu
Luton, UK
Capacity: 60,000 cars a year.
Output 45,000 in 1994

Suzuki
Linares, Spain
Capacity: 60,000 cars a year.
Produces compact sport/utility vehicles.

Nissan
Barcelona, Spain
Capacity: 135,000 cars a year.
Output 104,000 in 1994.

Toyota
Burnaston near Derby, UK
Capacity: 200,000 cars a year.
£700m investment.

Deeside, Clwyd
Capacity: 200,000 engines a year. First production September 1992.
£140m investment

Suzuki
Esztergom, north of Budapest, Hungary
Capacity: 50,000 cars a year.

Mitsubishi Motors/Volvo
Born, Netherlands
Capacity: 200,000 cars a year.

Japanese companies have set up a number of manufacturing plants to produce cars for the European market. This has been for two reasons. Firstly, to save on transport costs from Japan. Secondly, and more importantly, to avoid quotas (a form of import control) on the number of cars that Japan can export to the European market.

London in gridlock

Tuesday 10 December 1996. London traffic come to a standstill for eight hours. A lorry driver, ignoring height warning signs, had got his truck jammed into the entrance of the Blackwall Tunnel. Traffic all over the centre and east of the capital was brought to a standstill.

As frustrated drivers joined the queues, engines revved and tempers flared. And air pollution rose to dangerous levels. Many drivers trapped in the tunnel and other enclosed areas were forced to breath through scarves and handkerchiefs.

Cold, still air contributed to high levels of air pollution all over the UK in December. Traffic jams were reported on many major motorways. The M6 north of Birmingham was reported as the worst stretch of road for traffic chaos in the whole of Europe.

And still car sales continue to rise - but for how much longer in the UK?

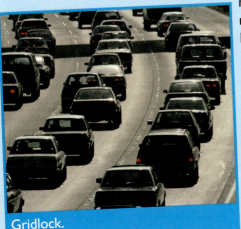

Gridlock.

What it takes to build a car

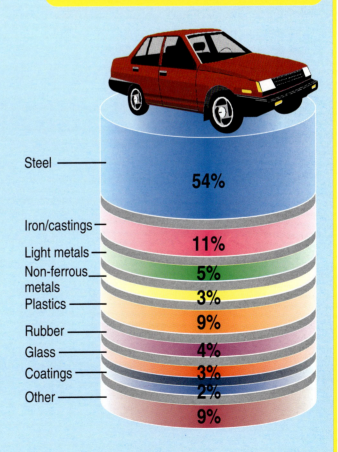

Steel	54%
Iron/castings	11%
Light metals	5%
Non-ferrous metals	3%
Plastics	9%
Rubber	4%
Glass	3%
Coatings	2%
Other	9%

Toyota comes to Derby

In 1989 the Japanese car manufacturer Toyota announced a £840m investment in the UK. It had chosen an old airfield at Burnaston, near Derby as its base in Europe. By 1992 Toyota was producing its first cars at the Derby site. 1,750 jobs were created at two factories at Derby and Deeside, North Wales, where the engines are built. By 1999, Toyota will make 200,000 cars per year at Burnaston with about 75% being exported to mainland Europe. Over 3,000 people will be employed.

So why did Toyota choose Derby when it had a long list of other sites in Britain and the rest of Europe from which to choose?

- The 580 acre site is alongside the main A38 trunk road which offers good links to the M1 and the M6. Derbyshire County Council promised to spend £12 million upgrading the A38 and other local routes.

- The local area offered a large, skilled workforce. The decline of British Railways Engineering and the coal industry had released a large pool of labour which Toyota could employ and train.

- Components were readily available from suppliers in the West Midlands which has a long association with the motor industry. Dunlop tyres, Lucas batteries and lights and car upholstery firms are all located nearby.

Note: although some companies receive government grants to fund their 'inward investment', Toyota chose to 'go it alone' and fund the investment themselves.

The location of Toyota's Burnaston plant

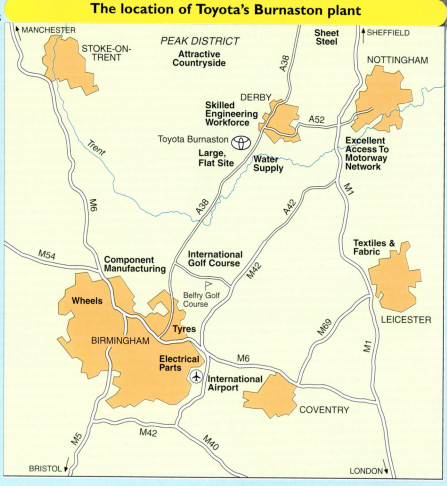

Ordnance Survey map of the Toyota site at Burnaston, Derby (Grid reference 290305)

Study the map and then carry out the following tasks:

1 Make a tracing paper overlay of the map extract. Draw around the main urban areas and industrial sites; mark on the Toyota site and the main lines of transport.

2 On, or around, your land use map add notes to illustrate why Toyota might have chosen this 'greenfield' site. (Consider, for example, transport links, workforce and suitable land.)

3 How might the following people have reacted to the building of the car plant:
   - an unemployed skilled engineer (ex Rolls Royce worker) living at Normanton, Derby (345338)
   - the owners of land which was purchased for the car plant
   - residents of Etwall village (270318) who now benefit from the improved trunk roads.

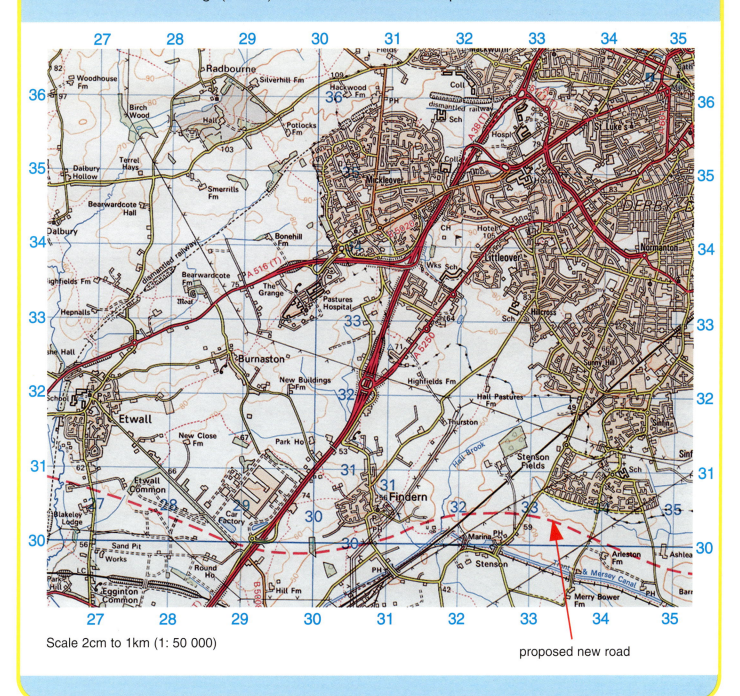

Scale 2cm to 1km (1: 50 000)

proposed new road

STAGE 3:   How to attract new industry?

We have seen that many of today's major industries have a very wide choice of locations. When Toyota decided to build a car plant in Europe, it could have chosen dozens of places. But it chose Burnaston near Derby! What are the factors that company bosses consider when choosing a site? How do local authorities and national governments try to attract this inward investment? In the resources that follow, these questions are answered.

Particular information is provided on Wales which has been very successful in attracting Japanese and other foreign companies.

Using the resources provided, make a list of the attractions and selling points being made. Put them into separate categories, for example; financial; social; transport; workforce; raw materials; environment.

On an outline map of Wales, mark on the areas which have attracted new industry. Make notes or labels that explain why the industries chose those locations. Compare your map with other students' and add any factors you might have missed.

The changing face of Wales

Sheep farming and coal mines? Not any more. Many people still have that image of Wales. A country with coal mining in 'The Valleys' and sheep farming in the hills of mid and North Wales. But nearly all the deep mines in Wales have closed and only 27,000 people in the country now work full-time in farming. The other Welsh giant, steel making, has undergone massive investment and restructuring. The old iron and steel works in North East Wales have closed with the exception of a rolling mill at Shotton. In the south, old works have closed and only two major modern integrated works remain - at Margam near Port Talbot and Llanwern near Newport, now the most efficient works in Europe. So whilst farming, mining and steel making have declined and their jobs have been lost, what has taken their place? Since 1990 the Welsh economy has enjoyed great success in attracting foreign and UK investment. About 20% of all foreign investment in the UK goes to Wales with more than 40 Japanese firms now located there. Sony, Toyota, Hitachi, Aiwa and Matsushita head the list and have now been joined by the Korean giant LG.

These new industries have largely been attracted to the M4 corridor in South Wales with a smaller number of firms attracted to North East Wales. Major road building, including the new second Severn Bridge, and the redevelopment of the Cardiff and Swansea docks have helped to transform the image of the area.

The Welsh Development Agency

- Established 1976.
- Aims to regenerate industry in Wales.
- Land Reclamation: 8,000 hectares of derelict industrial land turned into recreational land or modern industrial estates.
- Programme for the Valleys to help regenerate housing and jobs on the old coalfields.
- Enterprise zones at Swansea, Milford Haven and Delyn (N Wales). Attracts new investment by reduced taxes and government control for 10 years.
- **Assisted Area** status for much of Wales. This means that firms can apply for government grants to help finance their investment.

Advice, Training and Information and Ideas for Industry.

Industrial Estates and Business Parks. Purpose built Factories. Advice on Locations.

WELSH DEVELOPMENT AGENCY (WDA)

Tourism. Environmental Improvement. Rural Industries.

Advertising. Selling Wales to potential investors.

Urban Regeneration. Housing Schemes. Marinas - Swansea/Cardiff. Reclaims Derelict Land.

Why Wales?

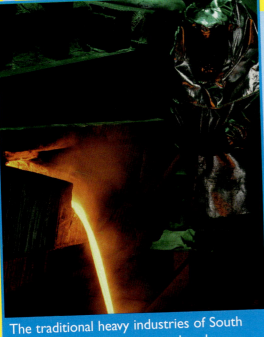

The traditional heavy industries of South Wales are gradually being replaced.

- University and Research Centres - the University of Wales in Cardiff, Swansea, Bangor and Aberystwyth.
- Reliable supplies of water and electricity.
- Excellent motorway connections via the M4 and Severn bridge.
- Cardiff International Airport.
- Close to supply of raw materials - coal/steel and components.
- A pleasant rural environment. Good beaches and golf courses.
- A large, skilled workforce and low wage rates.
- Cheap, ready prepared industrial sites.
- Close to large markets within Wales, the Midlands and the South East of England.
- Good port facilities.
- Government grants and tax incentives from the WDA and the European Coal and Steel Community.

Korea brings 20,000 jobs to Wales

The biggest overseas involvement ever to come to Europe has gone to Wales. The giant South Korean electrical and petrochemical conglomerate LG is investing £1.7bn in two new microchip and TV component plants. These will employ over 6,000 people on the Imperial Park site at Newport. More than 12,000 additional jobs will be created in factories supplying components and thousands more in construction. This is known as the **multiplier effect**.

LG is one of Korea's largest multinationals, alongside DAEWOO, HYUNDAI and SAMSUNG. These firms are looking to sell more in Europe by setting up factories here.

And why did LG choose Newport in Wales? A major reason is that this is an Assisted Area, ie an old industrial area with high unemployment where the government offers financial help to attract new industry. It is claimed that the government has given £200 million in grants to LG - or about £30,000 for every job.

Other reasons given by LG are:

- Low labour costs. Wage rates are relatively low in Wales, even compared with South Korea (which only 20 years ago was regarded as a poor, developing country).
- A free 250 acre site at Imperial Park with all services laid on - water, electricity, sewage, access roads.
- Lower taxes than elsewhere in Europe.
- Legislation which controls union activities in the UK.
- Closeness to London and the South East via the old and the new Severn Bridges.
- The English language is taught in Korean schools.
- Because the UK is a member of the European Union, goods produced in this country have free access to the whole EU market.

The Multiplier Effect

For every job at the LG factories in Newport at least one other, probably two, will be created within the region. The factories will buy their component parts from established or new factories in the area. New roads and houses will be built, transport companies will benefit and there will be a boost in retail sales in the surrounding region.

An industrial success story

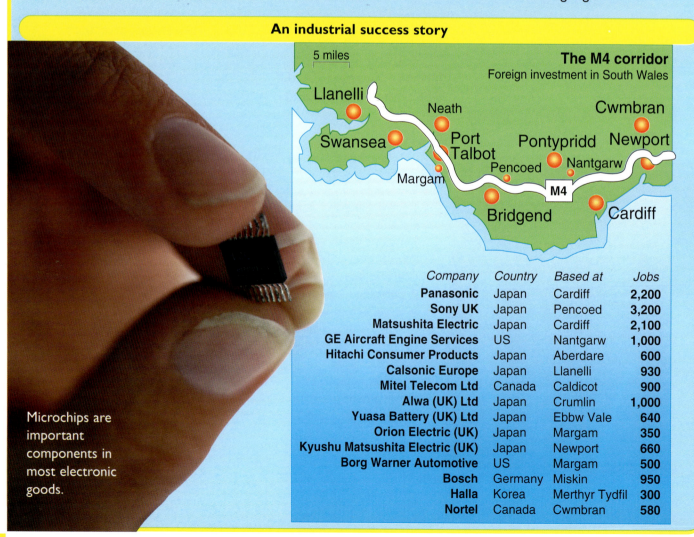

Microchips are important components in most electronic goods.

The M4 corridor
Foreign investment in South Wales

Company	Country	Based at	Jobs
Panasonic	Japan	Cardiff	2,200
Sony UK	Japan	Pencoed	3,200
Matsushita Electric	Japan	Cardiff	2,100
GE Aircraft Engine Services	US	Nantgarw	1,000
Hitachi Consumer Products	Japan	Aberdare	600
Calsonic Europe	Japan	Llanelli	930
Mitel Telecom Ltd	Canada	Caldicot	900
Alwa (UK) Ltd	Japan	Crumlin	1,000
Yuasa Battery (UK) Ltd	Japan	Ebbw Vale	640
Orion Electric (UK)	Japan	Margam	350
Kyushu Matsushita Electric (UK)	Japan	Newport	660
Borg Warner Automotive	US	Margam	500
Bosch	Germany	Miskin	950
Halla	Korea	Merthyr Tydfil	300
Nortel	Canada	Cwmbran	580

Development in Rural Wales

The Development Board for Rural Wales aims to attract new industry to mid-Wales by emphasising several features:

- a quality of life that is not possible in busy urban areas
- ready built industrial premises
- low rents and purchasing prices for new industrial premises
- a workforce that is educated and committed.

By emphasising modern methods of communication such as teleconferencing and teleworking (based on computers and modems), the Development Board seeks to overcome any reluctance to set up in a remote rural area.

Your millennium opportunity...

Innovation factories... instant occupation...

From only £1.95 per sq ft
(£21 per sq m)*

High Lifestyle Wales

Come and see Southern Snowdonia for yourself. Breathe the air. Eat the food. Test the comfort of our justifiably famous country hotels. And when you've a feel for the place, take in a possible business location or three with our Mike Rees.

In fact, he'll happily help you put your itinerary together and provide money saving introductions to most of the nicest places to recoup after the rigours of your fact finding mission.

We promise: no hard sell. No boring presentation. Just all the help you need to decide that your most inspiring prospects are here in Southern Snowdonia.

If you think Southern Snowdonia means splendid isolation, think again. In fact, it's one of the hubs of the British Business Park - the unique network of more than 50 business locations which connects the 100s of thriving Mid Wales companies. Each benefiting from the well developed business infrastructure offering support in training, research, product development, marketing and exporting.

Rural Wales
THE BRITISH BUSINESS PARK

TO SEE FOR YOURSELF, PHONE MICHAEL REES ON 0800 269300 OR POST OR FAX THE REPLY PAID CARD

Development Board for Rural Wales, Ladywell House, Newtown, Powys SY16 1JB

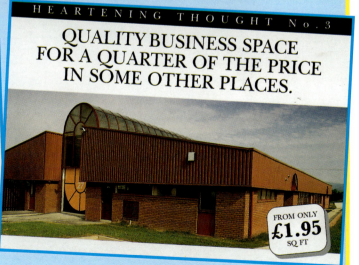

HEARTENING THOUGHT No. 3

QUALITY BUSINESS SPACE FOR A QUARTER OF THE PRICE IN SOME OTHER PLACES.

FROM ONLY
£1.95
SQ FT

STAGE 4: Finding a site

When a large company decides to establish a new factory, it usually finds a vacant area of land and builds its own plant. However, more and more companies are choosing to set up on an established **Industrial Estate**. This can save time and money, particularly if they can move into a ready built factory.

In the resources which follow, you will find information on Industrial Estates. Make notes, in the form of a spider or star diagram, that describes what an Industrial Estate is and why companies find them attractive sites.

Industrial Estates

Business Park, Industrial Estate, Science Park. A complicated mixture of terms but all have common features. They are areas of land with purpose-built premises and a well developed **infrastructure**. That is they are laid out with all the services (transport, water, power, communications) a new business would want. Whilst an Industrial Estate is most likely to be dominated by manufacturing industry it will also have some services. Business and Science Parks may have a greater emphasis on office premises, distribution depots and retail outlets.

Today most towns and cities have industrial estates. Some are very small scale whilst others cover huge areas of land. Often these estates are built on old industrial land, eg Trafford Park in Manchester, whilst others are established on **greenfield sites** on the edge of town and alongside major roads. The most successful are alongside motorways.

Swansea Enterprise Park

This contains a mixture of industrial, business and retail premises. It is landscaped with a lake as a central feature. The road layout is designed so that all the premises have fast access to the M4 motorway. The new premises are purpose built so they can accommodate modern machinery and production techniques.

What are Industrial Estates like?

Industrial Estates and Business Parks share several common features:

- They are devoted entirely to business and commerce. There are no residential areas mixed in with the industries.
- They have purpose built factories for sale or rent - or plots of land available for building.
- The estate is usually landscaped and made to look as attractive as possible.
- The estate is often built adjacent to open space so that there is room for expansion in the future.
- All services are laid on, ie sewerage, water supply, electricity, street lighting, gas and service roads.
- All factory sites have adequate car parking and loading areas.
- The estate is usually built close to a motorway intersection or a major road to provide good access for workers and for goods vehicles delivering components and distributing finished goods.

The estate will often have a mixture of **light industries** including electrical goods, textiles, food processing, warehouses and offices and may also have retail outlets such as DIY centres and kitchen and bathroom showrooms.

A typical new office and workshop complex on an industrial estate

Note the greenfield site. The surrounding tarmac area allows easy parking and road access for vehicles.

Food processing is a typical activity found on industrial estates. Nearness to consumers and good communications are vital.

Modern warehousing requires a large floor area and quick road access to motorways.

STAGE 5: Review

The previous Stages have contained information on new industries and reasons for their location. In this Review Stage you are asked to study and then recommend, in a short report, a location for a major new factory in the UK. The factory will be for Siemens, a large German multinational company which, in the mid 1990s, wished to expand its micro-electronics production.

Work individually or in pairs, but you must complete your own copy of the final report. Your report will need to include information on all the factors affecting location. It will be judged on the quality of your comments and the reasons you give for your final decision.

To keep things simple, you are given information on only three sites, from which you must choose one. Take all the factors into account, including the likely availability of labour (you are provided with details of local unemployment levels).

By 1995, Siemens narrowed its choice to one of three sites:

    A Dover, in South East England

    B Bolton, in North West England

    C Longbenton, in North East England

You are asked to set out the strengths and weaknesses of each location and then to recommend one of them as the site for the factory - together with your reasons. Sketch maps, with added notes would help your report.

Siemens

Siemens is one of Germany's largest companies. It is the biggest manufacturer of mobile phones in Europe and a world leader in the production of microchips. The company has interests in the development of all kinds of industrial technology including gas turbines for generating electricity, electrical machinery for water treatment plants, communication systems for all types of industry and electronic components for the car industry.

Siemens has been looking for a site for a major investment, a huge new semi-conductor factory where it will manufacturer 'wafers' for the microchip industry in the UK and Europe. These are Siemen's requirements in order of importance:

LAND: A 150 acre site available now. It needs to be flat and ready for building to begin.

MARKETS: The 'wafers' will sell all over the world but a good local market would be an attraction, eg local companies making TVs, computers, cars etc.

TRANSPORT: The wafers will be shipped all over the UK, across to Europe and the world. Siemens needs access to the UK motorway system, to ports and to an international airport, ideally all within an hour's drive.

LABOUR: Siemens intends to employ 2,000 workers. It will need a good supply of well qualified and skilled workers, preferably with a background in electrical, engineering and computer technology. It will also be looking for low wage costs.

TRAINING: The company will offer its own training but would like to work with a local university to provide training and Research and Development (R&D) facilities.

FINANCIAL ASSISTANCE: Siemens is investing over £1 billion in this development. The firm will expect some government incentives in terms of grants, rent-free property, tax concessions etc which are found in Assisted Areas.

Siemens makes a huge range of electronic goods. It is Europe's biggest manufacturer of mobile phones.

UK Assisted Areas

UK Transport Links

LONGBENTON

Location: Longbenton is situated in the North East region of England, just north of the conurbations of Tyne and Wear and Teesside.

Transportation Services: Longbenton is well served by a modern transportation network. The main north-south road links are provided by the A1(M) motorway. There are two international airports in the area, at Newcastle and Teesside. The main ports are Tyne, Tees and Sunderland; they offer roll-on, roll-off services to the rest of Europe, including Scandinavia, and container services to the rest of Europe, the Middle East, Africa and India. Newcastle lies on the main east coast railway line to London.

In an Assisted Area: Yes - companies locating in Longbenton have access to Regional Selective Assistance, the most generous kind of grant on offer in the UK.

Other Kinds of Financial Assistance on Offer: The Council can offer incoming companies grants to help offset such costs as property taxes (or rates) and site improvements, among other costs. Financial support to help pay for training is also available from the Tyne and Wear Training and Enterprise Council.

Available Sites Include: There is a wide range of serviced sites in sizes from one acre up to about 170 acres. All are within easy reach of major highways.

Prices: Average price per acre vacant serviced land £50,000; Rent per sq ft existing factories £3.00.

Weekly earnings: men £330 (89% of national average); women £243 (90% national average)

Unemployment rate (July 1995): 10.5%

Major Industries in the Area: Include electronics, electrical components, plastics, automotive components and clothing.

Major Companies in the Area: 3M, Black & Decker, Caterpillar from the US; Fujitsu Microelectronics, Sanyo, NSK from Japan; Philips Components, Electrolux from mainland Europe; Thorn Lighting from the UK.

Local universities: Durham, Newcastle, Northumbria, Sunderland,

Advice and Assistance: Companies interested in locating in Tyne and Wear can use the free services of the Council, which include a 'one-stop shop' covering advice on property, housing, education, meeting local suppliers, etc.

Summary: Tyne and Wear, with its east coast location, offers inward investors an ideal location from which to serve both the UK market and the rest of Europe. It provides a competitive environment in terms of land and property, and its workforce has a proven reputation for both quality and adaptability. Its appeal is best endorsed by the fact that it is currently home to more than 120 overseas companies from 21 countries.

DOVER

Location: The district of Dover is in the county of Kent, on the south coast of England, just 21 miles from France. London is one hour away by either rail or motorway. The main towns in the district are Dover, Deal and Sandwich.

Transportation Services: As the largest ferry port in Europe, Dover offers excellent cross-Channel links from the UK to the rest of Europe. There are up to 75 ferry and hovercraft departures each day. The port is literally ten minutes from flagship locations such as the White Cliffs Business Park (see below). The Channel Tunnel, offering direct rail-links to the rest of Europe, is ten miles from the port. For passengers, there is also an International Rail Station serving the Tunnel, which is 20 miles away. The district is connected to the M25 orbital motorway (that circles London) by the A20 and M20. It therefore has good links with areas throughout the UK.

In an Assisted Area: Yes - Companies locating in Dover have access to Regional Selective Assistance.

Other Kinds of Financial Assistance on Offer: Training Grants; and Innovation Grants.

Available Sites Include: Prime greenfield sites are at White Cliffs Business Park (195 acres) and Millenium Park (100 acres). Smaller sites ranging from two to eight acres are also available.

Prices: Average price per acre vacant serviced land £100,000; Rent per sq ft existing factories £5.50.

Weekly earnings: men £389 (104% of national average); women £276 (103% of national average)

Unemployment rate (July 1995): 8.2%

Major Industries in the Area: Include transportation and logistics, measuring instrumentation, box making and pharmaceuticals.

Major Companies in the Area: Pfizer Pharmaceuticals, employing some 2,500 people; London Fancy Box, employing 250, and P&O European Ferries.

Local universities: Canterbury.

Advice and Assistance: The Dover District Council will help incoming companies with key-worker housing. It also offers a full relocation service from the District Council's Economic Development Manager.

Summary: Dover's strategic location - in the part of the UK closest to the rest of Europe - is supported by excellent grants packages, the availability of skilled labour, low costs and superb communications. In addition, the district is a refreshingly different place to live: it is largely rural, has delightful villages, a magnificent coast line, and world-class golf courses.

Dover is the UK's busiest ferry port for passengers and containers. Its biggest advantage as an industrial location is its position.

BOLTON

Location: Bolton is located in the metropolitan area of Greater Manchester. It is a major centre in its own right. It lies about 12 miles northwest of Manchester city centre.

Transportation Services: Bolton has excellent road connections. A dual carriageway from the town centre provides a direct link with the M61, M62, M63 and M6 motorways. Because of its proximity to Manchester, Bolton also has excellent rail and air connections. It is, for example, close to the main north-south rail line that runs up the west coast of Britain. Nearby is Manchester's International Airport - used by more than 50 airlines serving some 180 destinations in the UK, the rest of Europe and elsewhere around the world. In addition, the Royal Seaforth container terminal is only one hour away. East coast ports are easily reached via the M62. And there is a terminal serving the Channel Tunnel at nearby Trafford Park.

In an Assisted Area: Yes. Companies locating in Bolton can apply for some of the UK's best grants.

Other Kinds of Financial Assistance on Offer: In addition to Regional Selective Assistance, companies locating in Bolton have access to training and employment grants. They can also receive enterprise grants for building refurbishment, security and environmental works.

Available Sites Include: Wingates Industrial Park, The Valley, Platinum Park, Rivington Parkway, Red Moss and Chequerbent. Two sites over 150 acres are available.

Prices: Average price per acre vacant serviced land £90,000; Rent per sq ft existing factories £4.00

Weekly earnings: men £352 (94 % of national average); women £252 (94% of national average)

Unemployment rate (July 1995): 7.5%

Major Industries in the Area: Textiles, food, all types of engineering, plastics, warehousing and distribution.

Major Companies in the Area: Warburtons Bakery, Robert Watson Constructional Steelwork, Riva, British Aerospace, Vernacare, Bernsteins, Fort Sterling, Ingersoll Rand, Edbro Metal Box, Sanderson Fabrics, Vantona.

Local universities: Manchester, Manchester Metropolitan, Salford, UMIST.

Advice and Assistance: Bolton Metropolitan Council can help incoming companies in all major areas including finding key-worker housing and obtaining introductions to local suppliers. It also offers an 'after-care' service for companies once they have invested in Bolton.

Summary: Bolton is located close to the West Pennine Moors, and about one hour's drive from the Lake District (the UK's largest national park), so it offers a quick transfer from an urban to a rural environment. It also has 14 golf courses in its immediate area, good schooling, and an award-winning town centre. Altogether, Bolton is able to offer a complete work environment. It has excellent motorway connections, a skilled and reliable workforce, pleasant environment, excellent shopping facilities and quality housing.

Glossary

All the terms listed below are explained in this Enquiry. In each case, write a definition of the term and, if possible, give at least one example.

Globalisation	**Research and Development**
Light industry	**Investment**
Footloose industry	**Assisted Area**
Automated production	**Development Agency**
Greenfield site	**Multiplier effect**
Electronics industry	**Industrial Estate**
Pharmaceutical industry	**Business Park**
Aerospace industry	**Infrastructure**
Multinational / Transnational company	

12 Global trade

During the past hundred years, the world appears to have shrunk. **Not literally!** - but in terms of people's attitudes and perceptions. Improved methods of transport and communication mean that information and finance can travel around the world almost instantaneously. Journeys that once took weeks can now be completed in hours.

With this revolution in transport and communication has come a massive increase in trade between countries.

In this Enquiry we shall look at different aspects of international trade. You will be asked to consider whether the increase in world trade leads to a fairer distribution (ie sharing) of wealth or helps to widen the gap between rich and poor.

questions to consider

1 Why do countries trade with each other?
2 What is the pattern of trade in goods around the world?
3 Who benefits from world trade?

key ideas

Global trade has increased rapidly in the second half of the twentieth century. The world is moving towards a system of *free trade*. This means that there are fewer controls on imports and exports. If countries specialise in producing goods and services at which they are relatively efficient, and then trading, all countries can benefit. This idea is known as *comparative advantage*. It is an important reason behind the setting up of the European Union where there is free trade between members. As the European countries specialise and trade then overall output rises and prosperity increases.

On a world scale, trade is dominated by a small number of *multinational* or *transnational* corporations.

Trade can be divided into *goods* and *services*. Goods include food, fuel, raw materials and manufactured goods. Services include tourism, transport and finance. This Enquiry focuses on the trade in goods.

activities

Using information on this and the facing page:

Briefly suggest two reasons why international trade is increasing.

Make a list of 10 goods which you know are imported into this country. Share your list with other students and classify them - into raw materials (which are used to make other goods) and consumer goods (which are the finished goods). Are there any goods which do not fit into these two categories? Make a separate list of them.

GLOBAL TRADE

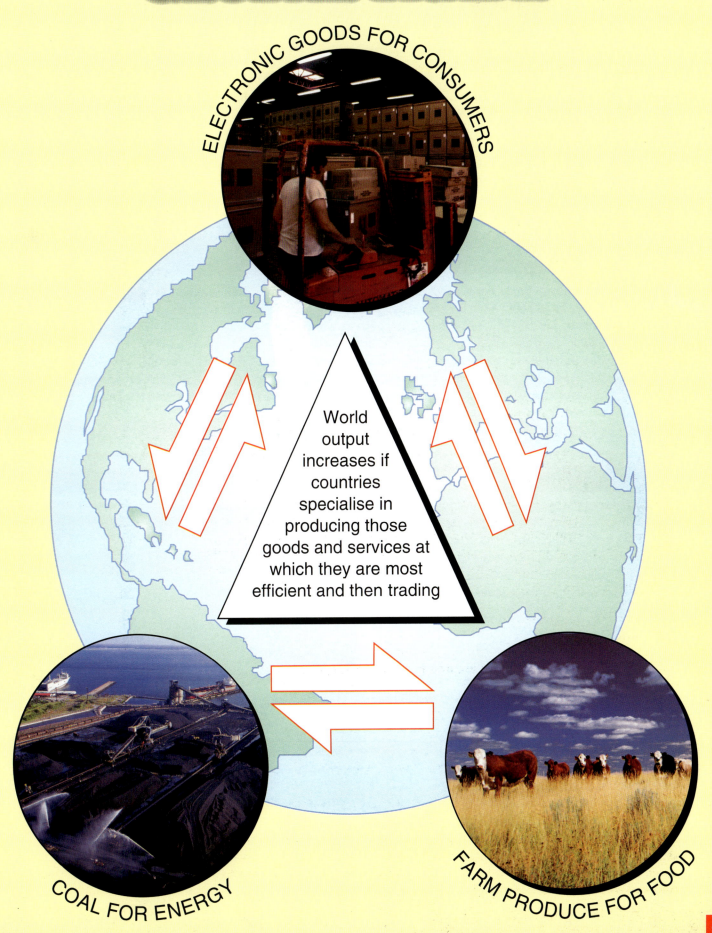

ELECTRONIC GOODS FOR CONSUMERS

World output increases if countries specialise in producing those goods and services at which they are most efficient and then trading

COAL FOR ENERGY

FARM PRODUCE FOR FOOD

Enquiry

How does trade affect different countries?

In this Enquiry you will investigate aspects of trade. Why does it take place, where does it take place and how does it affect different people and countries? During the past 50 years there has been a huge rise in world trade. Most of this rise has been between the richer, economically developed countries. However, a group of newly industrialising countries (NICs) - mostly in south and east Asia - have, in recent years, also become important world traders.

Over the same period, the pattern of UK trade has changed. At one time the pattern was for the UK to import food and raw materials and to export manufactured goods. Now, most of the UK's imports are manufactured goods.

This Enquiry consists of a number of statistical exercises. These will provide you with a portfolio of information about international trade. The outcome will be a comparison between the trade of two countries.

STAGE 1: Why is trade necessary?

Trade occurs because people need goods that they cannot efficiently produce themselves. If everyone was self sufficient, there would be no need for any trade. But this would be very inefficient and it is much better if people, regions and countries specialise. They produce the goods at which they are efficient and then trade these for goods they do not produce themselves.

In the resources which follow, you are given information on why different countries trade and how world trade has developed over the past 200 years. Some of the theories which explain why trade takes place are outlined. Use the information to produce a poster, diagram or illustration which shows the reasons why countries trade.

Complete the tasks that are included after the table of UK imports.

Why does international trade take place?

International trade happens when one country **imports** (ie buys) goods from another country or when it **exports** (ie sells) goods to other countries. Why does Britain, for instance, need to do this? Why cannot we produce all of the food, fuel and products we need? One reason is that certain foods are very difficult to grow in this country. Bananas can be grown in greenhouses but this is very expensive. They can be grown more cheaply in the Caribbean and imported to this country.

In manufacturing Britain has become very efficient at making some items, but less successful with others. So we sell the products we are good at making to other countries and use the money we make from this to buy products which we do not or cannot make ourselves. For example, Britain exports a large amount of armaments around the world and imports lots of clothing and footwear, including most of the training shoes you can buy in the shops.

Britain has developed experience and influence in banking and insurance so many countries buy these services from companies in this country.

International trade occurs because countries specialise in making particular goods. It would be very wasteful if every country tried to be self-sufficient.

Comparative Advantage

So why do some countries produce certain foods or manufactured goods more cheaply or efficiently than others? Some economists say that this is because of **comparative advantage**. This means that one country is relatively more efficient than another. So its goods are very competitive and can be sold around the world.

A good example of this was seen in South Korea during the 1970s when it became a world leader in the production of clothing. This was mainly based on its advantage of having a very plentiful and cheap supply of labour. But as South Korea became richer, wages rose and it lost its advantage. Today the clothing factories have moved to countries like Indonesia and the Philippines where wages are still low.

Another example of comparative advantage can be seen in the banana trade. Fertile soils, a favourable climate and investment by US multinational companies have made the countries of Central America the most efficient producers of bananas in the world. Colombia and Costa Rica are at the forefront. The reason that the UK imports most of its bananas from the West Indies is that we allow these bananas into the country without the import taxes which are charged on other countries' bananas. We do this because we have a historical link with the West Indies which used to be part of the British Empire, and also because the bananas are grown and exported from these islands by a British company, Geest, which controls much of the banana production.

It makes sense for Europe to import bananas from Central America and the Caribbean rather than grow them more expensively in greenhouses.

UK Imports (at 1994 prices)				
	1985		1994	
	£ millions	%	£ millions	%
Food and live animals	12,264	9.4	12,379	8.2
Beverages and tobacco	1,870	1.4	2,265	1.5
Raw materials (eg timber, metal ore, fibres)	7,465	5.7	5,549	3.7
Fuels	16,228	12.5	6,153	4.1
(Petroleum)	(12,652)	(9.7)	(4,843)	(3.2)
Animal and vegetable oils	910	0.7	537	0.3
Chemicals	10,538	8.1	14,614	9.7
Manufactured goods				
(eg paper, textiles, metals)	21,656	16.7	24,538	16.3
Machinery and transport	41,102	31.7	60,917	40.6
(Road vehicles)	(10,366)	(8.0)	(16,275)	(10.8)
Other manufactured goods				
(eg clothing, scientific goods)	15,527	12.0	21,916	14.6
Other goods	1,829	1.4	990	0.6

Draw bar charts to show which imports have risen or fallen between 1985 and 1994.

Draw pie charts / divided circles to show each category of import as a % of the total in 1985 and 1994.

Briefly describe what your graphs show.

A brief history of world trade

1500 European countries conquer and colonise the 'New World', ie the Americas.

Sugar, tobacco, spices and cotton, gold and silver are taken from these colonies.

Fortunes are made in the slave trade. 10-12 million slaves shipped from Africa to Brazil, the Caribbean and North America.

Britain establishes a new trading Empire.

This Empire provides food and raw materials for a rapidly growing population and a captive market for the manufactured goods, eg cotton and locomotives, being made in the new industrial cities. Although conditions are often harsh in the mills and factories, Britain, France and Germany become very rich in this unequal trade.

Two World Wars, the Wall St. Crash of the New York Stock Market and the great Depression damage world trade. Countries set up import 'barriers' to protect their own industries. This causes International Trade to decline even further and there is mass unemployment.

During the Second World War the USA becomes the world's greatest economy.

1914

1945

After the Second World War the USA and its allies set up the General Agreement on Trade and Tariffs (GATT) and the Organisation of Economic Cooperation and Development (OECD). These richest countries agree to encourage trade by removing trade barriers. Industry booms in the USA and, later, in Germany and Japan. The 'oil crises' of the early 1970s damage trade for a while but rapid advances in transport and communications 'shrink' the world and trade quickly recovers.

A deep recession in the late 1980s damages western industry. Japan suffers trade deficits for the first time. Multinational Corporations become more dominant as controls are loosened. The new 'Tiger Economies' of the Far East emerge - Hong Kong, Singapore, Taiwan and South Korea. They particularly specialise in electronics and consumer goods.

The Soviet Union breaks up.

1985

1996

STAGE 2:   Patterns of world trade

In Stage 1, we saw why countries trade with each other. In the resources which follow, we look at the pattern of trade, in other words which countries trade with each other. You will see that the world is divided into **economic or trading blocs**. These are groups of countries, such as the European Union, which have joined together to promote trade between themselves.

Each statistical resource that follows is accompanied by an activity which you are asked to complete. Then make a brief note on trading blocs, giving examples and membership details. Then explain what is meant by free trade and protection.

The major world trading blocs

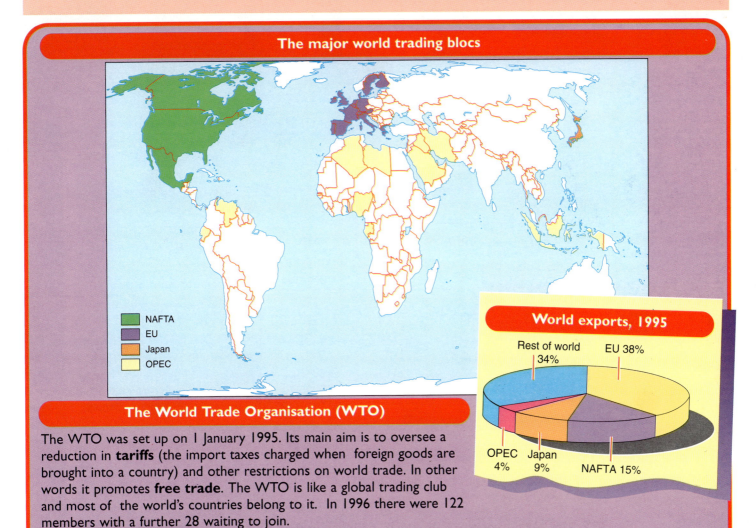

NAFTA
EU
Japan
OPEC

World exports, 1995

Rest of world 34%
EU 38%
OPEC 4%
Japan 9%
NAFTA 15%

The World Trade Organisation (WTO)

The WTO was set up on 1 January 1995. Its main aim is to oversee a reduction in **tariffs** (the import taxes charged when foreign goods are brought into a country) and other restrictions on world trade. In other words it promotes **free trade**. The WTO is like a global trading club and most of the world's countries belong to it. In 1996 there were 122 members with a further 28 waiting to join.
(Formerly the WTO was known as GATT - the General Agreement on Trade and Tariffs.)

NAFTA: The North American Free Trade Agreement. NAFTA was signed in 1994 by the three countries of North America. Two of these - the USA and Canada - are amongst the richest countries in the world whilst Mexico, the third, is much poorer. It is hoped that companies will set up in Mexico to take advantage of the low wages but with free access to the huge North American market. This should then increase prosperity in Mexico and reduce the number of migrants to the USA.

The European Union (EU): 15 countries in Europe are now members of the EU. Within the EU, trade barriers are disappearing rapidly with the aim of completely free trade. This is accompanied in Europe by agreements on wages and trade unions (the 'social chapter') and moves towards a single currency - the Euro. NAFTA and the EU are the world's most powerful trade groups but others exist in South East Asia, Africa and South America.
List the countries which belong to the: a) NAFTA , b) EU, c) OPEC.
Which major areas of the world are not included in these 3 trading blocs?

The growth in world trade

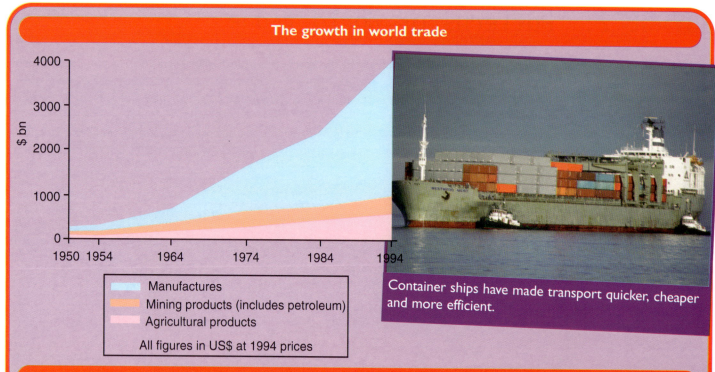

Legend:
- Manufactures
- Mining products (includes petroleum)
- Agricultural products

All figures in US$ at 1994 prices

Container ships have made transport quicker, cheaper and more efficient.

The exports of selected countries (US $ billion)

Year	EU	USA	Japan	All developing countries
1970	112	41	20	39
1975	333	112	68	126
1980	566	175	110	218
1985	650	208	185	341
1990	1390	400	314	674
1993	1410	440	358	822

Describe what has happened to world exports since 1950. Compare the increases of agricultural, mining and manufactured products. Give 3 examples of each type of product. Using the figures provided, draw a similar line graph to show the growth of exports from the European Union, the USA, Japan and the developing countries. Briefly note what your graph shows.

The biggest trading nations in 1995 (US $ billion)

Country	Imports	Exports	Surplus / deficit
USA	770.8	583.9	
Germany	441.7	506.4	64.7
Japan	335.9	443	
France	272.4	284.5	
UK	264.7	239.9	
Italy	203.4	232.5	
Netherlands	177.5	197.5	
Hong Kong	197.6	173.8	-25.8
Canada	171.8	192.1	
Belgium / Lux.	152.5	165.9	
China	132.0	148.8	

Note: You can see that Germany exported $64.7 billion worth of goods more than it imported. Hong Kong exported $25.8 billion less than it imported. So Germany had a **surplus** (more exports than imports) and Hong Kong had a **deficit** (more imports than exports) .

Describe, in global terms, where the world's biggest trading nations are located. In other words, which continents are they in, and are they in the North or South?
Draw bar charts to show the imports and exports of each country. Calculate the surplus or deficit for each country. Germany and Hong Kong have been done for you. Which are the other countries which had a deficit?

World trade in oil (1994)

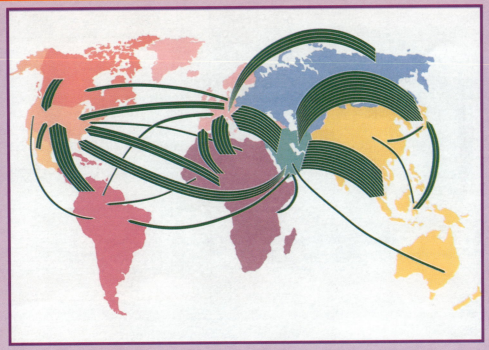

Which areas import the most crude oil? Suggest possible reasons for this.
On an outline world map draw arrows to show the 6 biggest oil flows in 1994.

Free trade or protection?

During the twentieth century there has been a big difference of opinion about which of these two approaches to trade is the best:

Protectionism: This means a country protects its own industries against foreign imports. If a country cannot produce something efficiently itself, eg bananas in the UK, then it might allow them to be freely imported. But if a country fears that its own industries are threatened it might impose import duties (taxes on imports called **tariffs**), or **quotas** (numerical controls on imports). This has happened with the import of Japanese cars to the USA and the EU.

Free trade: Free traders argue that everyone is better off if there are no tariffs or quotas. This allows countries to concentrate on what they produce efficiently and sell it all over the world. Prices should be lower and more goods will be sold. A problem with this approach is that less efficient businesses might be driven out of existence, especially in the poorer countries of the world.

Some groups of countries, like the EU, have grouped together and agreed to free trade between themselves whilst putting up trade barriers to protect their industries from outside. Since the 1940s the world has moved towards freer trade; a system which greatly benefits the richer countries and the multinational corporations.

Oil output in the USA is not sufficient to meet the country's needs. 'Gas guzzling' cars, air conditioned homes and massive industries all need to use imported fuel.

Free trade or fair trade

The idea of free trade sounds attractive at first, but sometimes the truth is more complicated. With no controls, the rich countries and companies make most of the profits and fix the rules to suit themselves.

The World Trade Organisation talks about 'fair trade' as well as Free Trade but it remains to be seen how this will be achieved. Fair trade includes:

- guaranteed prices and markets for primary products like tea and coffee
- minimum wages and acceptable working conditions for employees anywhere in the world
- protection of the environment in areas where foreign multinationals own land, operate mines or establish factories
- the removal of trade barriers aimed at developing countries (eg higher taxes on imported processed coffee than on raw coffee beans)
- the removal of subsidies for farmers in the developed world (because these give the farmers in the developed countries an unfair advantage)
- control over the activities of multinational companies (so they have to consider the interests of the poorer countries).

Consumers in developed countries can encourage fair trade by buying goods which have been fairly traded and by putting pressure on retailers to stock these goods.

The examples of tea and coffee

Tea and coffee are the world's favourite drinks. Huge quantities of both are grown and consumed every year. Tea and coffee are primary products which are grown mostly in developing countries (LEDCs) in the tropics. Brazil is the world's biggest coffee exporter and India is the world's biggest tea exporter. The main markets for these drinks are in the richer developed countries, the USA, Western Europe and Japan. Much of the production of the tea and coffee is controlled by multinational companies such as Nestle. Such companies own large plantations in the developing countries and they control the growth, harvesting, processing, packaging, transport and retailing of the product.

So who benefits from this trade? The developing country gains jobs with low wages and some income from tax. The developed countries, like the UK and the USA, get the tea and coffee they want at a low cost whilst most of the profits go to companies in their own countries.

Why don't the developing countries charge more for the tea and coffee? Because if they try, the big companies simply go and buy their tea and coffee from other countries willing to sell it at the prices offered because they are desperate for trade.

Why don't the developing countries establish their own companies to process and sell the tea and coffee and so get the profits themselves? Two reasons - firstly, they don't have the capital to establish large scale industries which could compete with the multinationals and, secondly, the developed countries charge import duties on processed tea and coffee, making it too expensive.

UK coffee market shares

Kraft General Foods 20.8%

Brooke Bond Foods 3.5%

Nestlé 57.9%

Supermarket 'own label' 14.8%

Other brands 3%

Note: most of the supermarket 'own brand' coffee is produced by the multinationals under contract.

Who gets what from a jar of coffee?

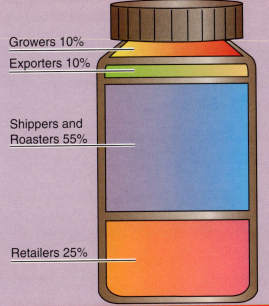

Growers 10%
Exporters 10%
Shippers and Roasters 55%
Retailers 25%

Kenya is the biggest tea producer in Africa.

Most coffee is grown in low income countries. Most is consumed in high income countries.

Growers and drinkers

Tea

Top 10 producers (000 tonnes)			Top 10 consumers (000 tonnes)		
1	India	760	1	India	560
2	China	600	2	China	400
3	Sri Lanka	230	3	Russia	270
4	Kenya	210	4	UK	160
5	Indonesia	130	5	Japan	130
6	Turkey	130	6	Pakistan	120
7	Japan	90	7	Turkey	90
8	Russia	70	8	USA	80
9	Bangladesh	50	9	Iran	80
10	Argentina	50	10	Egypt	61

Coffee

Top 10 producers (000 tonnes)			Top 10 consumers (000 tonnes)		
1	Brazil	1,800	1	USA	1,080
2	Colombia	670	2	Germany	600
3	Indonesia	440	3	Brazil	540
4	Mexico	250	4	Japan	360
5	Guatemala	210	5	France	330
6	Vietnam	180	6	Italy	290
7	India	170	7	Spain	170
8	Ethiopia	170	8	Netherlands	160
9	Kenya	140	9	UK	160
10	Ivory Coast	130	10	Indonesia	140

Make some brief notes on the world trade in tea and coffee. Mark on a world map the main producers and consumers. Suggest reasons for the pattern of trade.

If the producers of tea and coffee could agree between themselves to fix the world price higher than at present, who would benefit and who would suffer?

The information in Stage 2 shows that world trade is dominated by the economic giants of North America, Japan and Europe. But the economies of some of Japan's neighbours in the **Pacific Rim** (countries situated around the Pacific Ocean) and in Latin America (South and Central America) have grown rapidly in the past three decades. They are becoming important traders on a world scale. They are sometimes described as **Newly Industrialising Countries** (NICs).

The information which follows looks in more detail at these rapidly growing countries. Complete the task that is outlined at the end of the statistical data on NICs.

South Korea is used as an example to illustrate the growth of NICs. Use the information to write a short summary that describes and explains South Korea's growth. Draw up a table or graph to show the number of Samsung factories in each continent. Suggest some possible reasons for this distribution.

List the factors which have contributed to South Korea's growth.

Economic growth and the NICs

Country	Growth rates (average % per year)			Total % growth (65 - 94)
	1965 - 1980	1980 - 1990	1990 - 1994	
Brazil	8.8	2.7	2.2	170
Mexico	6.5	1.0	2.5	120
Hong Kong	8.6	6.9	5.7	227
South Korea	9.6	9.4	6.6	271
Singapore	10.1	6.4	8.3	257
Taiwan	9.4	6.5	7.0	241
Indonesia	8.0	6.1	7.6	219
Malaysia	7.3	5.2	8.4	204
Philippines	5.9	1.0	1.6	106
Thailand	7.2	7.6	8.2	225
Developed world (includes the UK)	3.6	3.2	1.7	95

On a world map, draw bar charts (similar to the one shown below) next to the 6 NICs at the top of the table to show their growth during each period shown.

Which country has had the fastest growth rate since 1965?

Economic growth in Brazil

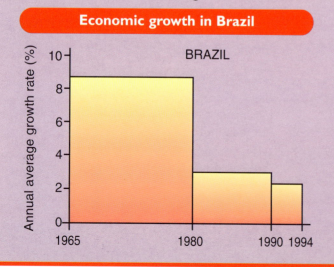

Singapore. The NICs of the Pacific Rim have outperformed those in Latin America.

NICs

In the 1960s and 1970s a group of countries with high economic growth rates were called Newly Industrialising Countries (NICs). They included South Korea, Taiwan, Singapore, Hong Kong, Mexico and Brazil. Since then, the East Asian NICs have outperformed their Latin American counterparts and are now sometimes called the four **Tiger** economies. Singapore and Hong Kong have developed so fast that their GNP per person is higher than the United Kingdom's. Other East Asian countries have also enjoyed rapid growth rates. They include Malaysia, Indonesia, Thailand and the south eastern part of China.

Latin American NICs have not grown as fast as those in East Asia. Unstable governments and high inflation have made them less attractive to foreign investors.

The Tiger Economies

Two factors help explain the success of the East Asian Tiger economies. Firstly, they have specialised in producing manufactured goods for export. Secondly, they have attracted multinational companies such as Sony and Nike to set up factories making electronic goods, textiles and footwear. Low labour costs, compared with Europe and North America, have been an important factor in attracting the multinational companies. However, countries like Singapore, South Korea and Taiwan have made great efforts to train and educate their work forces and this in turn has attracted 'high tech' industries. The result is that there now exists a rapidly growing middle income group of people who enjoy standards of living similar to those in the USA and European Union.

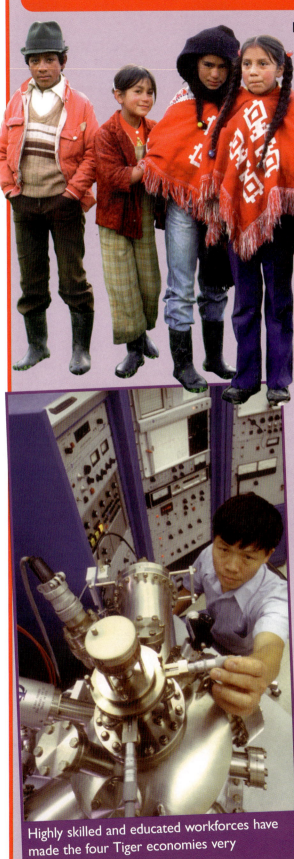

Highly skilled and educated workforces have made the four Tiger economies very competitive.

South Korea brings 20,000 jobs to Wales

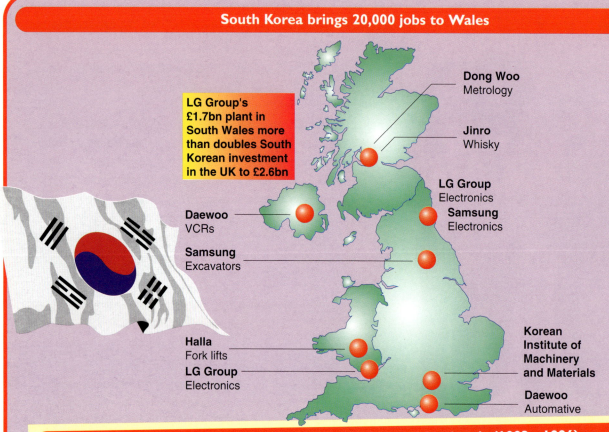

LG Group's £1.7bn plant in South Wales more than doubles South Korean investment in the UK to £2.6bn

Dong Woo Metrology

Jinro Whisky

LG Group Electronics

Samsung Electronics

Daewoo VCRs

Samsung Excavators

Halla Fork lifts

LG Group Electronics

Korean Institute of Machinery and Materials

Daewoo Automative

World wide investment by South Korean multinationals (1993 - 1996)

	US $	What they make
Daewoo	1.4 bn	Cars, steel, VCRs, ships
Hyundai	1.2 bn	Cars, excavators, trucks, trains, ships, chemicals
Samsung	1.2 bn	Household electronic goods
LG (Lucky Goldstar)	1.0 bn	Computers, electronic goods, ships

The Samsung Corporation worldwide

Europe HQ LONDON

America HQ NEW YORK

China HQ BEIJING

Japan HQ TOKYO

SE Asia HQ SINGAPORE

• Production centres

Fact File: SOUTH KOREA

South Korea Factfile

Area: 99,392 sq km
Population (1994): 44,453,000
Currency: Won
Language: Korean
Main towns and population:

Seoul (capital)	10,613,000
Pusan	3,798,000
Taegu	2,229,000
Inchon	1,818,000
Kwangju	1,139,000
Taejon	1,049,000

Main products

Computers and semiconductors
Cars
Ships
Clothing and footwear
Petrochemicals
Steel

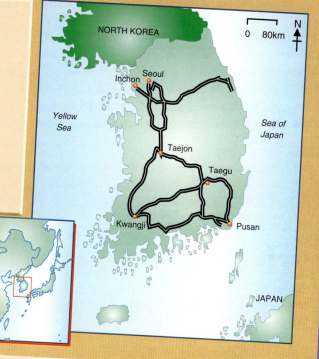

Growth

In the 1960 and 70s, South Korea's military rulers encouraged large business organisations to set up manufacturing industry. They attracted large amounts of US and Japanese investment. Cheap and plentiful labour produced clothing and footwear at first, later steel and ships. Most recently cars and electronic goods have been the economic driving force.

As Korean wages have risen, some large companies have started to manufacture abroad. Many of the clothing and footwear businesses have moved to the lower wage economies of South East Asia. Korean cars and electronic goods are produced around the world - usually in locations near their markets.

During 1996 and 1997, South Korean workers have started to campaign for higher wages and the right to set up independent trade unions. Having worked hard to create their 'economic miracle' they now want a greater share in some of the benefits.

	1993	1996
GDP per capita ($)	7,670	10,873
Foreign debt ($bn)	74,619	93,428

GDP Share 1993

Agriculture	6%
Manufacturing	44%
Services	50%

Trading partners (%)

	Exports	Imports
USA	21.4	21.1
Japan	14.1	24.8
Hong Kong	8.3	6.1
China	6.5	5.3
Germany	4.5	5.0

Trade ($m) 1996

Imports	143,100
Exports	133,250
balance	- 9,850

Main exports

Transport equipment
Electrical goods
Footwear
Textiles

STAGE 4: The multinationals

In South Korea some of the most successful companies such as Samsung and Daewoo have expanded overseas and built factories all over the world. A company which produces in more than one country is known as a **multinational** or **transnational corporation** (**MNC** or **TNC**). Most multinationals originate in the richer countries - as we might expect. However, as the South Korean examples show, more and more multinationals come from the south and east Asian tiger economies.

Some of the largest multinationals are so big that their annual revenue is greater than the entire value of national output in many less economically developed countries.

The resources which follow contain information on multinational companies and how they affect world trade. You are asked to complete the activities that are included with the resource material. Then, in note form, list the 'good' and 'bad' points about multinationals. In other words, describe the advantages and disadvantages of multinationals from the point of view of people (consumers and workers) living in different countries.

Unilever

A range of Unilever products.

This is an example of a multinational company. It is based in the UK and the Netherlands but has operations in 90 other countries. In addition, its goods are sold in a further 70 countries.

Unilever produces consumer goods such as detergents, frozen foods, margarine and hair shampoo. Its brand names include Persil, Omo (washing powders) Colman's mustard, Flora margarine, Brooke Bond tea, Oxo gravy, Wall's ice cream, Calvin Klein cosmetics and Organics hair shampoo.

The company's activities provide a good example of **globalisation**. This is a process in which events, decisions and activities in one part of the world affect other parts of the globe. Unilever buys its raw materials from many different countries and then makes the finished products in the countries that form the biggest markets. Its brand names are known the world over. For example, the Organics range of hair shampoo is sold in over 40 countries. It was developed by Unilever scientists working in Paris and Bangkok and was first launched in Thailand in 1993. Now it has worldwide sales of over £200 million per year. Similarly, Omo is sold in 50 countries. In some countries the detergent is slightly adapted to meet local needs but the strength of the brand image brings world wide sales of over £750 million per year.

Unilever's worldwide operations

Countries in which Unilever has operations

Additional countries in which Unilever's brands are sold

GNP for selected countries ($m) 1994

	Total GNP
Bangladesh	26,164
Colombia	67,266
Denmark	146,076
Egypt	42,923
Finland	97,961
Greece	77,721
Indonesia	174,640
Malaysia	70,626
Singapore	68,949
Tanzania	3,378
UK	1,017,306

Note: GNP is the total annual value of all goods and services produced.

Where products are made by MNCs.

Mobile phone – South Korea; Reebok trainers – Indonesia; Philishave – Netherlands; BASF tape – Germany; Sony radio – Japan; Minolta camera – China; Black & Decker iron – Singapore

Annual revenue of the world's largest multinational corporations (MNCs) 1994 (Top 10 and others selected)

Rank	Company Name	1994 Revenues US $m	Country in which head office is located
1	Mitsubishi	184,365	Japan
2	Mitsui	181,518	Japan
3	Itochu	169,164	Japan
4	General Motors	168,828	USA
5	Sumitomo	167,530	Japan
6	Marubeni	161,057	Japan
7	Ford Motor	137,137	USA
8	Toyota Motor	111,052	Japan
9	Exxon	110,009	USA
10	Royal Dutch Shell	109,833	Neth / UK
14	Daimler-Benz	72,256	Germany
15	IBM	71,940	USA
20	Nissan Motor	62,568	Japan
21	Volkswagen	61,489	Germany
22	Siemens	60,673	Germany
23	British Petroleum	56,981	UK
27	Daewoo	51,215	South Korea
31	Unilever	49,738	Neth / UK
32	Nestle	47,780	Switzerland
33	Sony	47,581	Japan
34	Fiat	46,467	Italy
55	Samsung	35,060	South Korea
59	Procter & Gamble	33,434	USA
67	BMW	32,199	Germany
77	Pepsico	30,421	USA

Which of the countries listed has a GNP which is larger than the revenue of the largest MNC?
Which countries have a lower GNP than any of the top 10 MNC's revenue?
Describe the distribution of the MNC's headquarters. In other words, are they in richer or poorer countries and which continents are they in?

The top 10 multinational employers 1994		
Company	Overseas employees (000s)	Total employees (000s)
General Motors (US)	270	756
Pepsico (US)	128	423
Siemens (Germany)	153	403
Daimler-Benz (Germany)	82	367
Ford(US)	180	333
Unilever (Neth/UK)	187	294
Fiat (Italy)	67	288
IBM (US)	130	256
Matsushita Electric (Japan)	98	254
Volkswagen (Germany)	103	253

Multinationals and their employees

As you would expect, large multinationals employ many thousands of people. Some in the 'home' country and often many others abroad. Indeed one of the main reasons for setting up factories abroad is to find cheaper labour so that production costs can be kept low and profits high.

Multinationals and their 'host' countries

There is no doubt that developing countries welcome investment by MNCs. They need the jobs, the income created through taxes and the infrastructure (eg roads and power lines), which is often provided. Sometimes, though, the jobs provided are not as good as they could be. Some MNCs contract-out work to local companies so they are not directly responsible for poor wages or unsafe conditions.

Some MNCs have been criticised by the International Labour Organisation (ILO) and charities, such as Oxfam, for their treatment of workers. This is particularly true of clothing and footwear companies producing in the poorest countries such as Haiti, Indonesia and Bangladesh. Extremely low wages are sometimes paid, combined with long hours of forced overtime and few employment rights. For example, trade unions are rarely allowed to operate.

The case of the 1984 Union Carbide plant explosion at Bhopal in India was one of the worst examples of poor safety standards. An estimated 7,000 have died and many more were injured due to the poisonous gas that escaped. Lower safety standards than those allowed in the USA were blamed, together with the fact that the plant was built next to a very densely populated urban area.

For and against multinationals

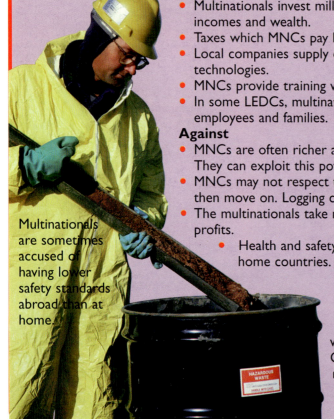

Multinationals are sometimes accused of having lower safety standards abroad than at home.

For

- Multinationals invest millions of pounds in foreign countries, raising employment, incomes and wealth.
- Taxes which MNCs pay help provide social and other government services.
- Local companies supply components and parts and so learn new processes and technologies.
- MNCs provide training which has a beneficial effect through the economy.
- In some LEDCs, multinationals provide housing, health care and education for their employees and families.

Against

- MNCs are often richer and more powerful than the government of the 'host country'. They can exploit this power to treat their employees poorly.
- MNCs may not respect the natural environment. They strip a region of its resources then move on. Logging companies in South East Asia have been criticised for this.
- The multinationals take most of the money they make out of the 'host' country as profits.
- Health and safety standards, and pay, are far below the levels in the MNC's home countries.

You (or your family) buy clothes, footwear, electronic goods and food made by multinationals in less developed countries. List the advantages and disadvantages (for you and the workers involved) of a successful campaign to improve the MNC workers' wages and conditions.

Carry out a survey of your home to identify all the products made by MNCs. Try to locate where they are made and the 'home' country of the MNC. Prepare a display of the information.

STAGE 5: The arms trade

One aspect of world trade that has changed rapidly in recent years is the arms trade. The trade in weapons is an issue that causes concern to many people. Although it brings wealth to the companies which produce the arms, some low income countries spend more on weapons than on economic development.

Until 1990, the pattern of trade in arms was dominated by the 'Cold War' between the communist USSR and the capitalist USA. Since the collapse of communism and the break up of the USSR, the arms trade has become freer and less dominated by the two superpowers.

Read the information on the arms trade. Make brief notes on the trade. Who are the main suppliers and who are the main purchasers? Discuss whether the UK government should tighten or relax its controls on the arms trade. Make a list of points for and against the arms trade. Draw a poster or write a paragraph under one of the two following headings:

Stop the Arms Trade, or

Free Trade in Arms Now.

The main buyers and sellers of arms (1989 - 1995)

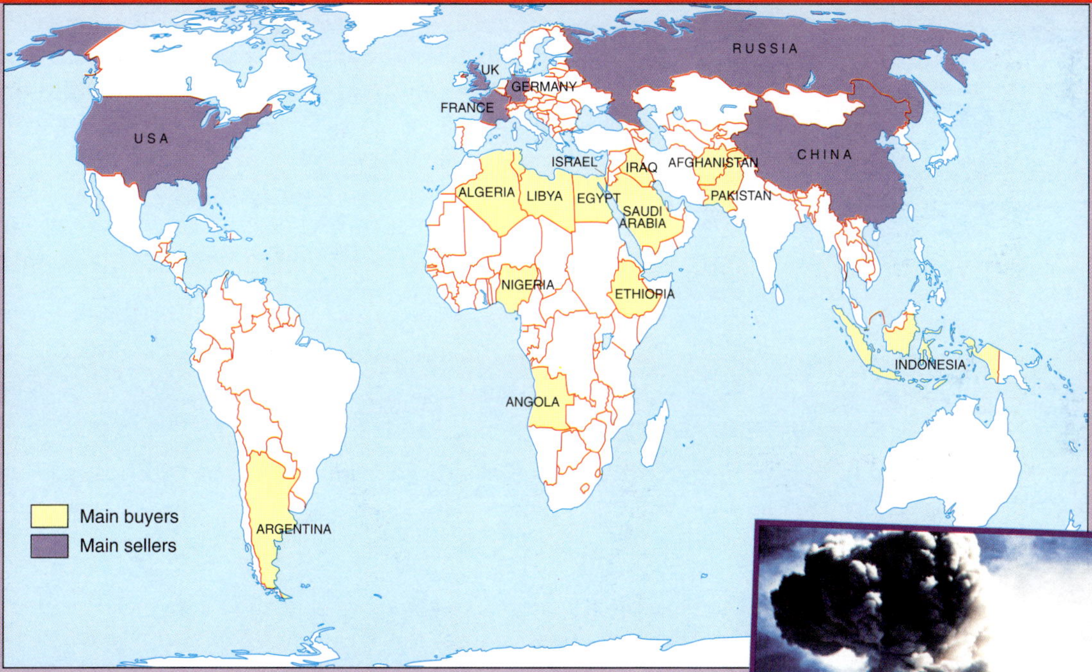

Main buyers
Main sellers

Why sell arms?
- to create and maintain employment in the arms industry
- to earn money from exports
- to maintain and support hi-tech industries which 'spin off' from the arms industry
- to gain influence in the countries which buy the arms
- to support regimes which are 'friendly'.

Why buy arms?
- to defend against enemies
- to fight and reclaim territory that has been occupied by another country
- to keep control over the population and prevent / suppress rebellions by opponents
- to maintain the power and prestige of the army.

Many people rely on the arms trade for their jobs. The irony is that we all hope the weapons will not be used.

Saudi deal breaks all records

In 1988 Britain signed its largest ever export deal. £10 billion worth of military equipment was sold to Saudi Arabia. Much of this will be paid for in oil. Tanks, warships, missiles and planes are all part of the deal. An estimated 200,000 jobs will be safeguarded by this one deal. Altogether, in the UK, 560,000 people have jobs related to the arms industry.

The Saudi government is extremely rich because of its massive oil reserves and exports. However, it feels threatened from abroad by Iraq in particular. It also wishes to ensure that there is no threat to its regime from internal instability or rebellion.

Indonesia promises not to use British aircraft in East Timor

There was outrage from peace campaigners when a deal was announced in 1992 to sell 24 British Aerospace (BAe) Hawk jets to Indonesia. The government in Indonesia has been much criticised, especially since its illegal 1975 invasion of East Timor. The invasion has been condemned by the United Nations and a leader of the East Timor resistance has been awarded the Nobel Peace Prize.

An estimated 250,000 people in East Timor have been killed or wounded by the Indonesian army in an effort to quell the resistance movement. The European Union has an agreement not to sell arms to countries with poor human rights records or internal conflicts. Indonesia has both, but the British government and BAe have stated that the jets will be used for training purposes only.

Princess Diana joins the campaign against landmines - January 1997

Diana, Princess of Wales joined the International Red Cross in urging a ban on the trade in anti-personnel mines during a visit to Angola. Around the world more than 2,000 civilians are killed or injured by landmines every month. Most of the casualties are in countries such as Angola which has experienced a long civil war. An estimated 110 million active mines are scattered in 64 different countries including Cambodia, Mozambique, Iran and Bosnia.

Most of the mines are manufactured in the developed countries of the world. Britain, like many countries, supports the ban but claims that until all the other producers agree, there is little point in one country stopping producing and exporting the mines.

The real cost of armaments

An issue that peace campaigners often raise is the real cost of weapons for people in poor countries. A large number of the world's population live below the poverty level. The price of one jet would pay for many health clinics, classrooms and roads in the poorer areas. On a world scale, just half a percent of total arms spending would buy all the equipment needed to make the poorest countries self sufficient in food. Fifty percent of one year's arms spending would pay for a 10 year programme to meet the food and health needs of the world's poorest twenty countries.

Is the arms trade wrong?

Some people argue that it is immoral for Britain to manufacture and sell any weapons at all. Others say that we should make only those weapons which are needed to defend Britain. Yet others believe that it is OK to sell weapons as long as we are sure that they will not be used to attack other countries or suppress the people of that country. And there are those who say that it is fine to sell as many weapons as possible so that Britain can make as much money as possible.

In the Gulf War, Britain needed weapons to help drive the Iraqi army out of Kuwait which it had attacked. Unfortunately, some of the weapons used by the Iraqis against British troops had been made in Britain!

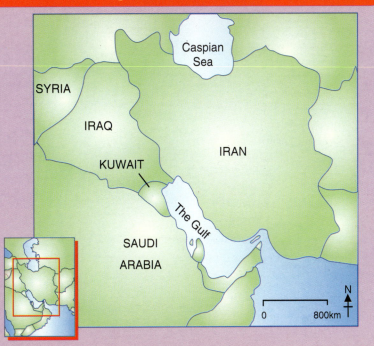

STAGE 6: Review

By now, you have a lot of information about world trade. On the next page is a Factfile containing information on the international trade of the UK. On page 229, there is a similar Factfile on South Korea. Write a short summary that compares the trade of the two countries.

If you have access to trade statistics for other countries (for example, the World Development Report), compile a Factfile for another country. It would be a good idea to work with a partner and one produce a Factfile for a less economically developed country and the other person produce a Factfile for a more economically developed country. Then compare these two countries.

Glossary

All the terms listed below are explained in this Enquiry. In each case, write a definition of the term and, if possible, give at least one example.

Goods	Protectionism
Services	Tariffs
Imports	Quotas
Exports	Pacific Rim
Comparative advantage	Newly Industrialising Countries
Economic (or trading) blocs	Tiger economies
Surplus	Multinational (Transnational) corporation
Deficit	Globalisation
Free trade	

Fact File: UK

UK Factfile

Area: 244,046 sq km
Population (1996): 57,800,000
Currency: Pound sterling
Language: English
Main cities and population of urban areas:

London (capital)	6,400,000
Glasgow	1,060,000
Birmingham	1,400,000
Manchester	1,670,000

Main products

Oil and gas
Machinery and transport
Chemicals
Processed foods
Steel
Textiles
Computers and electronic goods

Growth

In the nineteenth century Britain was known as 'the workshop of the world'. It dominated world trade. Its Industrial Revolution was based on abundant supplies of coal, profits from the colonies and the eighteenth century slave trade, and technological inventions.

During the twentieth century, Britain's traditional industries of coal, steel, textiles, engineering and shipbuilding declined relative to other countries. However, in the 1980s and 1990s, increased investment from Japan, Germany, USA and South Korea helped revitalise industry. Electronics and car making particularly benefited from the inward investment by foreign multinationals.

Trading partners (%)

	Exports	Imports
Germany	13.1	15.4
USA	12.4	12.3
France	9.6	10.8
Netherlands	6.7	7.1
Japan	2.1	6.2
Total EU	56.3	52.1

	1994	1996
GDP per capita ($)	18,000	18,900
Foreign debt ($bn)	-	-

GDP Share 1993

Agriculture	2%
Manufacturing	32%
Services	66%

Trade ($m) 1994

Imports	145,349
Exports	134,611
balance	- 10,738

Main exports

Machinery and transport equipment	41%
Chemicals	14%
Other manufactures	27%
Petroleum	6%
Food and drink	5%

Coursework Enquiry

Hypothesis 1: The existing system of international trade works against the interests of developing countries.

Hypothesis 2: The operations of multinational companies bring both costs and benefits to the countries where they set up their factories.

Choose **one** of the hypotheses for your investigation.

Hypothesis 1. You can approach this hypothesis in two ways. Either choose a developing country whose trade you can research (and about which you can find enough information). Or choose a primary commodity such as bananas, tea, copper or tin and investigate the world trade in that product.

Hypothesis 2. In this study you will investigate the activity of one multinational company. You could concentrate on one country in which it operates (including the UK), or consider its global pattern of operations. For example, it could be 'Coca-Cola world-wide' (or Unilever, or Ford etc.) or it could be 'Reebok in Indonesia'.

Method of enquiry and report writing

(Some ideas for a study on coffee are provided in italics.)

1 Explain your hypothesis - from what you already know, why do you think the statement might be true? As a starting point you might use a news story that features your chosen company, country or commodity. Set out the background to the story.
(What do you know about the trade in coffee?)

2 Decide what data you will need and how you will obtain it - for this research you will use secondary sources. In other words published information, for example from books, magazines or newspapers. It is much easier to base your study around information that you know is available rather than choose a topic and hope that you will find source material.
   Useful sources of information include the Internet, multinational companies (ask for their annual report), publications from the World Trade Organisation, the World Bank (World Development Report), Geographical magazines and digests, the New Internationalist magazine and 'Understanding Global Issues'. Browse through the Internet and your local reference library to see what is available before you finally decide which topic to research.
(World producers and consumers of coffee. How is coffee grown, processed and sold? Who controls these processes?)

3 Data collection - use the sources suggested above, atlases and textbooks.
(Embassies of producer countries; Coffee Council; Nestle; New Internationalist, July 1995; Oxfam.)

4 Record the data - select the information that is useful and relevant. Set out tables and draw graphs, maps and diagrams that will illustrate your report.
(Maps showing major producing and consuming countries. Prices of coffee beans and instant coffee.)

5 Analyse the data - describe what your tables, maps etc. tell you about your chosen topic.
(Who produces the coffee? Who profits from the production and trade?)

6 Interpret the data - decide whether or not the information supports or contradicts your hypothesis. Explain why.
(Does the coffee trade benefit the producers or consumers or both? Who gains the most?)

7 Evaluation and follow up - how successful was your study? What additional information would have been helpful? What was difficult about your study?
   As a follow up you might investigate how multinational companies and international trade affect your family. For instance, which companies and countries supply the food you eat, the clothes you wear and the household appliances that you use? On a much smaller scale it is interesting to investigate the component parts of a single bicycle. It will almost certainly be made by a multinational company and its parts (gears, frame, wheels, tyres, brakes etc.) will be produced in different countries.

Dedication

To Noel Cook for his encouragement and support
To Cath and the world's Puffins

Acknowledgements

Cover design Caroline Waring-Collins (Waring Collins Partnership)
Page design Caroline Waring-Collins
Graphics Elaine M Sumner
Cartography Stephen Ramsay Cartography (Pages: 21, 28, 35, 37, 44, 55, 56, 57, 73, 77, 83, 84, 105, 106, 110, 122, 145, 159, 164, 168, 180, 181, 202, 204, 206, 213, 221, 230, 233)
Advice Kenneth J. Youde B.Ed., Head of Geography, Ashton-on-Ribble High School, Preston
Reader Annemarie Work

Photograph credits

Frances Cook
Melissa Donnovan for her help with the Ingram Valley photographs
David Gray
Chris Pellant, Earth Science, Natural History and Landscape Photographer and Consultant
Robert Stainforth (who also provided fieldwork data)
Thomas Nettleship
Royal Geographical Society, London
Satellite Station, Dundee University
Toyota UK
Traidcraft Exchange (who also provided advice and support)
UNICEF/Heather Jarvis/1995
Agence France Presse; Hunting Aerofilms Ltd; [Images] 1997 PhotoDisc, Inc; Jefferson Air Photography; Panos Pictures; Rex Features; Still Pictures; Telegraph Colour Library; The Image Bank

Other data

Environment Agency. The data is supplied by the Hydrology Department of the Environment Agency - North East Region. Every effort has been made to check the accuracy of data supplied.
Meteorological Office: Crown Copyright. Reproduced by permission of the Controller of Her Majesty's Stationery Office
Ordnance Survey : © Crown Copyright 87508M
Sri Lanka Tourist Board: Annual Statistical Report

Text

HarperCollins Publishers for permission to use the extract from Gerald Durrell's 'My family and other animals'.
Every effort has been made to locate the copyright owners of material used in this book. Any omissions brought to the attention of the publisher are regretted and will be credited in subsequent printings.

British Library Cataloguing in Publication Data
A catalogue record for this book is available from the British Library
ISBN 1 873 929 684
Causeway Press Ltd, PO Box 13, 129 New Court Way, Ormskirk L39 5HP
First edition 1997
© Frances Cook, Helen Harris, Rachel Lofthouse, Jim Nettleship, Mel Rockett

Origination and layout by Caroline Waring-Collins (Waring Collins Partnership), Ormskirk, Lancashire.
Printed by Butler & Tanner Ltd, London and Frome